Starfishes, Serpent Stars, Sea Urchins and Sea Cucumbers of the Northeast

Starfishes, Serpent Stars, Sea Urchins and Sea Cucumbers of the Northeast

[formerly titled: ECHINODERMS OF CONNECTICUT]

Wesley Roswell Coe

DOVER PUBLICATIONS, INC.
NEW YORK

Published in Canada by General Publishing Company, Ltd., 30 Lesmill Road, Don Mills, Toronto, Ontario.

Published in the United Kingdom by Constable and Company, Ltd., 10 Orange Street, London WC 2.

This Dover edition, first published in 1972, is an unabridged republication of the work originally published by the State Geological and Natural History Survey, Hartford, Connecticut, in 1912 [Bulletin No. 19; State of Connecticut Public Document No. 47] under the title *Echinoderms of Connecticut*.

International Standard Book Number: 0-486-20949-0
Library of Congress Catalog Card Number: 76-185584

Manufactured in the United States of America
Dover Publications, Inc.
180 Varick Street
New York, N. Y. 10014

Preface

In the preparation of this report on the Echinoderms to be found in that part of Long Island Sound which borders the Connecticut coast line the writer has been influenced by two incentives: first, to aid in arousing an interest in the study of a most remarkable group of animals of common occurrence; and, second, to aid in the dissemination of useful knowledge of the structure and habits of the starfish — the deadliest of foes to the oyster, which constitutes one of the state's valuable natural resources, and with which a large and important industry is connected. With these ends in view, an attempt has been made to furnish such adequate illustrations of each of the species of Echinoderms found in the waters of the state as will facilitate their easy recognition, while the economic importance of the starfish has been as fully investigated as the conditions would permit.

Free use has been made of the many excellent treatises on the animals concerned; and grateful acknowledgment is made to Professor A. E. Verrill, of Yale University, for the use of specimens and photographs; to Professor Hubert Lyman Clark, of the Museum of Comparative Zoölogy, Harvard University, for the use of a number of the photographs which illustrate his excellent report on the Echinoderms of the Woods Hole region; and to Mr. Davenport Hooker and Mr. Stanley C. Ball, for assistance in the preparation of the illustrations.

In the bibliography at the end of the report will be found a list of some of the most useful publications on our native species of Echinoderms, including those on the economic importance of the starfish.

Contents

List of Illustrations

PLATES

FIGURES IN TEXT

INTRODUCTION

Among the natural objects most likely to attract the attention of the visitor to the seashore, particularly if the coast be somewhat rocky, are the five-pointed star-shaped creatures known as starfish. These are often found abundantly clinging to the rocks and seaweeds when the tide is low.

Among the seaweeds below the reach of the tides are somewhat similar animals, which have five slender, wriggling, snake-like arms, commonly known as serpent stars.

Just below the reach of the low tide also are other related animals, of rounded form, covered with coarse spines, which are commonly known as sea-urchins.

And, finally, when digging in the sand between tides, one may encounter pink or whitish worm-like animals, without distinct appendages to the body, but which nevertheless are able to cling to the hand; and these may belong to the group of animals known as sea-cucumbers, although our native species have little resemblance to the vegetable from which they take their popular name.

These four classes of animals occur abundantly along the shores of Long Island Sound, and careful search and critical examination will reveal the fact that there are several species of each.

All these animals belong to the division of the animal kingdom known to zoölogists as *Echinoderms;* for, although the starfish, sea-urchin, serpent star, and sea-cucumber may seem at first glance to differ from each other about as widely as any animals can, yet a careful study of their internal anatomy shows that all are built upon the same general structural plan. They furthermore have in general a similar embryology and life history. Hence the zoölogist and evolutionist believe that they are all related to each other, and have no hesitation in placing them together in a single group, or phylum, known technically as the *Echinodermata*.

These Echinoderms may be distinguished from other animals by certain well marked characters. As the name implies, the skin in most forms is provided with calcareous spines which project out from all sides of the body. Such spines are wanting however in the sea-cucumbers. Other calcareous plates are found in the skin. In most forms such plates unite to make a more or less complete shell which serves to protect the delicate internal organs. Most of them have likewise a distinct radial symmetry, in that similar parts are repeated around the central axis, as illustrated by the arms of the starfish or serpent star. With few exceptions they have a well developed water-vascular system, consisting of a circular canal around the mouth, with radial branches leading out to the periphery of the body. From the radial canals numerous branches lead out into the skin in certain regions of the body, and grow out into delicate tubular or finger-like projections.

These delicate projections are variously modified in different parts of the body for the performance of certain special functions. In all cases they serve as sense organs and as respiratory organs. If they have no further function they are known as tentacles; but in many cases they serve also in the locomotion of the body, and are variously known as tube-feet, sucker-feet, or pedicels.

These animals are found only in salt water, with the exception of one or more tropical species which are able to live in brackish water.

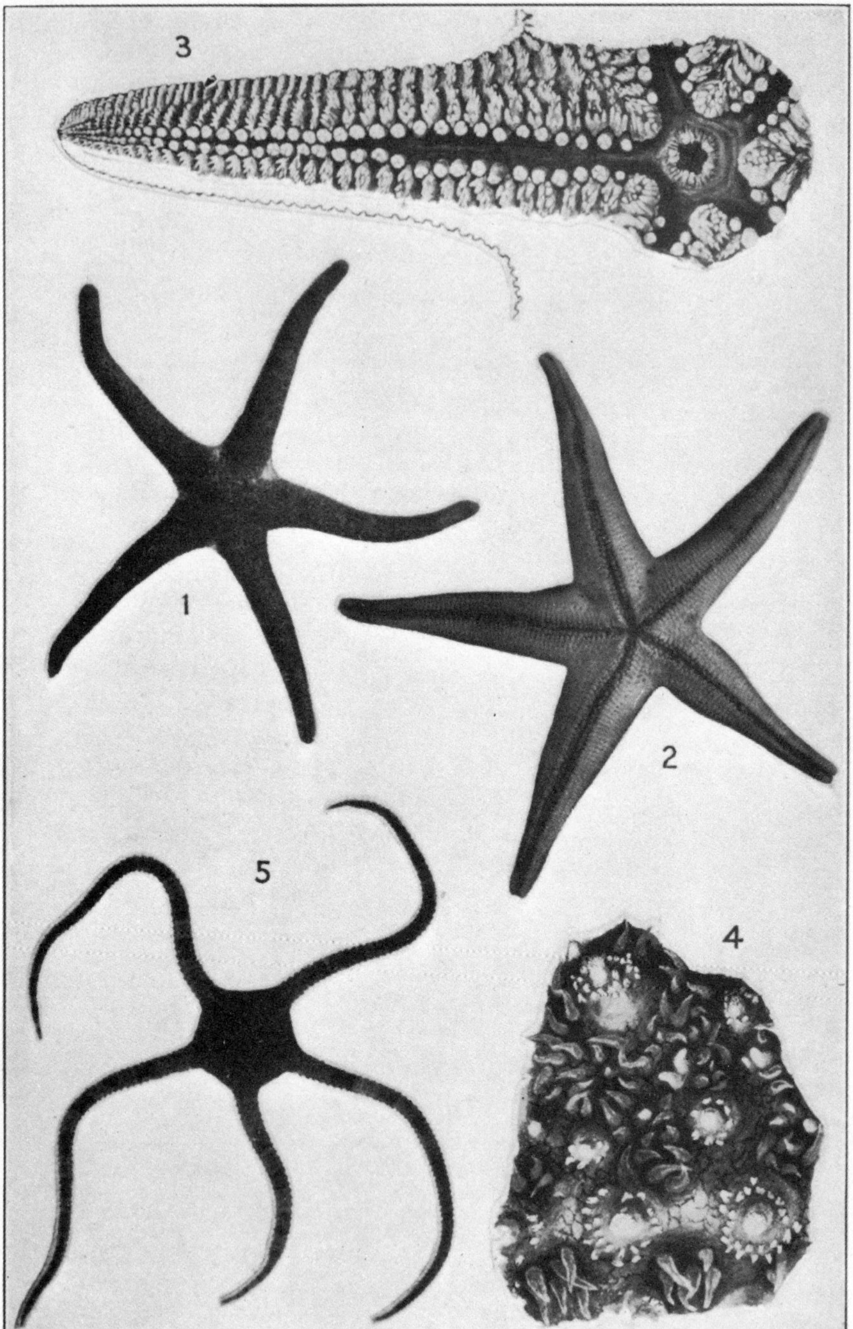

PLATE I. Figs. 1-3, *Henricia sanguinolenta;* Fig. 4, *Asterias forbesi;* Fig. 5, *Ophiura brevispina.*

CLASSIFICATION

All the Echinoderms of the world are divided by zoölogists into two sub-phyla, only one of which, the *Eleutherozoa,* is represented in our local fauna.*

The Eleutherozoa are divided into four classes, all of which have representatives living in Connecticut waters.

Class 1. *Asteroidea:* Starfishes.— The body consists of a central disk which passes without sharp demarcation into five or more symmetrically arranged arms or rays (Plate I). The large mouth occupies the center of the disk on the ventral, or oral surface, while the minute anus is situated opposite, that is, on the aboral side of the disk. The arms are without joints. A deep groove on the oral side of each arm is thickly covered with tube-feet which serve to move the animal from place to place.

In Long Island Sound occur four species of this class, each of which has five arms. They comprise the common starfish (*Asterias forbesi*), the northern starfish (*Asterias vulgaris*),

* The second sub-phylum, *Pelmatozoa,* which includes the *Crinoidea,* or Feather-stars, is represented in deep water as far north as off the coast of Maine by a species known as *Antedon dentatus,* but this has not been recorded from Long Island Sound. Farther south, and particularly in very deep water, occur other species, including some of the beautiful "sea lilies."

EXPLANATION OF PLATE I. Structure of Starfish and Ophiuran.

Fig. 1. *Henricia sanguinolenta.* Aboral surface of slender-armed specimen. (Natural size.)

Fig. 2. Same species. Oral surface of broad-armed specimen. (Natural size.)

Fig. 3. Same species. Portion of oral surface of disk and one arm, to show mouth, ambulacral groove, with the two rows of tube-feet, and spines. (Five times natural size.)

Fig. 4. *Asterias forbesi.* Small portion of aboral surface, showing the blunt spines surrounded by circles of pedicellariæ, and between the spines the closely-placed conical or tubular branchiæ. (Much enlarged.)

Fig. 5. *Ophiura brevispina.* Aboral surface. (Natural size.)

(Figs. 1, 2, and 5 from photographs loaned by the U. S. Fish Commission with permission of Dr. H. L. Clark.)

the slender-armed starfish (*Asterias tenera*) and the blood starfish (*Henricia sanguinolenta*).

Class 2. *Ophiuroidea:* Ophiurans, Serpent Stars, or Brittle Stars.— In these forms the body consists of a flattened, circular central disk, from the under side of which the usually slender, round arms, with a large number of movable joints, extend radially. These arms are sharply marked off from the disk (Plate I), and are provided with strong muscles, which enable them to be moved in nearly all directions and to be bent in sinuous or serpentine lines. By the movements of these arms the body is carried rapidly from place to place, and the flexibility of the arms enable the animal to hide away in a very narrow crevice. The mouth is situated as in the starfish, but there is no intestinal opening.

The ophiurans are represented in Long Island Sound by five species, only two of which occur abundantly. They comprise the following: the green serpent star (*Ophiura brevispina*), the daisy serpent star (*Ophiopholis aculeata*), the small gray serpent star (*Amphipholis squamata*), and the less common species, known as *Amphioplus abditus,* and *Amphioplus macilentus.*

Class 3. *Echinoidea:* Echinoids, or Sea-urchins.— Body nearly circular in outline, hemispherical, or flattened and discoid; without projecting arms. The calcareous plates in the skin form a compact encasement for the body except in the mouth region. The case is called the shell or, more properly, the test. The mouth is usually in the middle of the oral surface. The ambulacral areas, with numerous sucker-feet, in the more typical forms, extend around the body as meridians from the mouth nearly to the opposite pole. The spines are movably articulated with the plates on which they rest.

In Long Island Sound there are four species of echinoids, which differ so widely in general appearance as to be easily distinguished. The common purple sea-urchin (*Arbacia punctulata*) has a hemispherical test, with long, dark, reddish or purple spines. The green sea-urchin (*Strongylocentrotus dröbachiensis*) has shorter spines of a greenish color. The common sand-dollar (*Echinarachnius parma*) is very flat and covered with minute brownish spines. The key-hole urchin (*Mellita pentapora*) is likewise discoid, and covered with minute brownish spines; but is easily recognized by having five large openings, called lunules, which pass quite through the body from one surface to the other.

Class 4. *Holothuroidea:* Holothurians, or Sea-cucumbers.—The members of this group differ widely in general appearance from those of the other classes. Our native species all live beneath stones or burrow in the sand or mud. Their bodies are cylindrical and elongated, presenting a closer superficial resemblance to worms than to any of the other groups of Echinoderms. This worm-like appearance is greatly enhanced by the absence of any closely connected skeletal structures, and by the soft, bag-like, highly muscular skin. The skin has, however, isolated calcareous plates, often of such peculiar and definite shapes that the species are easily recognized by even a single plate. The mouth is situated at one end of the body, and is surrounded by a circle of branched tentacles. The anus is at the opposite end of the body. The animals move both by means of sucker-feet and by worm-like contractions of the body.

In Long Island Sound there are probably six species belonging to this class, although only four of these have as yet been recorded from Connecticut waters. The common *Thyone* (*Thyone briareus*) lies buried in soft mud in shallow water in great numbers in certain localities. Sometimes the animals have only the posterior end of the body projecting above the surface of the mud, but at other times they show also the anterior end with its circle of branched tentacles of black or purple color. The two other species of *Thyone, T. unisemita* and *T. scabra* are rare, and occur only in rather deep water.

Our fauna includes also two species with very slender bodies belonging to the genus *Synapta.* These are peculiar in having calcareous plates in the form of perfectly shaped anchors imbedded with other plates in the skin. *Synapta inhærens* is white in color, while *Synapta roseola* is pink or reddish. Both live in sandy localities between tides.

Finally, there is the beautiful sea-cucumber (*Cucumaria pulcherrima*), which is rarely seen alive, but specimens of which are sometimes thrown up on the beach by the waves following a storm. Little is known of its habits, but it is thought to live buried in the sand below low-water mark. It is short and thick, with firm body walls of a whitish color, and has peculiar plates in the skin.

The species enumerated above comprise all the Echinoderms known to occur in Long Island Sound, although the biological survey of these waters is far from complete. It is quite probable that future investigations will reveal the presence of additional species. The fauna of the eastern portion of the Sound includes more species than that of the western portion, because of its more varied physical conditions and its immediate connection with the Atlantic sea bottom.

Although the Echinoderms of this region are of no value commercially, they are of great economic importance because of the enormous injury which results from the depredations of the common starfish on the oyster beds. The members of the other classes feed mainly on seaweeds, minute organisms, and organic matter found upon the rocks and in the mud and sand between tides and covering the floor of the sea. They are therefore quite harmless to human interests, and are of some economic value in that they furnish a portion of the food supply of some of our edible fishes. Although covered with so firm an armor of defensive plates and spines, such species as the sand-dollar (*Echinarachnius parma*), the key-hole urchin (*Mellita pentapora*), and the species of serpent stars are nevertheless greedily devoured by the cod and other fishes. The holothurians with their less formidable plates are seized by various species of fishes, dragged from their burrows, and devoured. In their larval stages, too, the Echinoderms of all classes are eaten in enormous numbers by the surface fishes. It is however difficult to estimate the actual importance of these animals in respect to the human food supply.

The various groups of Echinoderms are likewise of great scientific interest, because of their remarkable anatomical structure, their peculiar modes of development, and their curious habits. No other group of the animal kingdom illustrates more clearly the modifications which the same organ-system can undergo in its adaptations to a great variety of functions. The evidence as to the relationships of the present-day Echinoderms, supported by an abundance of paleontological material, furnishes one of the best proofs of evolution.

Specimens of all classes of Echinoderms may be preserved in 80 per-cent alcohol. Formalin should not be used unless absolutely necessary, and should then be replaced by alcohol as soon as possible. Starfishes, serpent stars, and sea-urchins may be narcotized by adding magnesium sulphate (Epsom salts) to the sea water in which they are placed, and then killed by short immersion in fresh water or weak alcohol. They should then be placed in 50 to 70 per-cent alcohol for a few hours, and afterward preserved in 80 per-cent alcohol. For the preservation of holothurians with their tentacles expanded, it may be necessary to add the magnesium sulphate to the sea water very gradually. Otherwise the animals may contract vigorously.

Starfishes, ophiuroids, and echinoids make useful specimens when carefully dried. But all such specimens should be killed as indicated and placed for a day or two in strong alcohol. They should then be dried thoroughly by artificial heat. To prevent future injury from insect pests it is desirable to add a very minute quantity of corrosive sublimate to the alcohol in which the specimens to be dried are first preserved. It must be remembered that corrosive sublimate is a most powerful poison and must be handled with very great care.

The general scope of this report does not admit of an extended account of the internal structure of the various species included. For those who desire to make a thorough study of the organs of the body cavity of these animals the detailed descriptions given in various handbooks and laboratory guides in zoölogy are available. In the following chapters, therefore, will be given only such descriptions of the external features and internal organs as are necessary for an understanding of the relationship of the various groups, their natural history, including habits and life history of the various species, and for the accurate determination of each species.

Under the description of the species, only those characters which serve to distinguish our native species or have a direct bearing on their mode of life will be mentioned. More detailed anatomical descriptions of most of the species may be found in a report by Hubert Lyman Clark, entitled, "The Echinoderms of the Woods Hole Region," published in 1904, in the Bulletin of the United States Fish Commission for 1902. Many of the species are described, with interesting observations on their natural history, in Verrill and Smith's report on the "Invertebrate Animals of Vineyard Sound and Adjacent Waters," published in 1874, in the Report of the Commissioner of Fish and Fisheries. Other papers, which deal with particular species or groups of species, are mentioned under the separate species, while some of the more general treatises on the Echinoderms are included in the Bibliography at the end of this report.

PLATE II. Skeleton of Starfish, *Asterias forbesi.* (Natural size.)

CLASS I. ASTEROIDEA

The members of this group, popularly known as starfishes, are represented in Long Island Sound by three, or perhaps four, species, one of which is the commonest echinoderm found along the coast. They occur at all depths below low-water mark, and prefer rocky and shelly bottoms as a rule, because the mollusks which constitute a large part of their food are most abundant in such situations. They sometimes live on sandy, muddy, and gravelly bottoms, and are sometimes found adhering to piles of wharves. They are often found between tides, hiding beneath seaweeds or rocks in the tide pools, where they wait until the return of high water to resume their search for barnacles or mollusks adhering to the rocks.

All the species of starfish found in this region have normally five broad rays attached to a central disk (Plate II), although individual specimens having either more or less than this number are by no means rare. This variation in the number of rays is of much interest and is discussed in some detail on page 43. Along the northern coasts of New England lives a species which regularly has six rays and another which has ten; and in other parts of the world are species in which the number of rays increases with the growth of the animal until there may be as many as twenty to forty, or even more.

The starfishes creep slowly along by means of numerous tube-feet arranged in longitudinal rows on the under side of the rays. They can twist the flexible body and arms into almost every conceivable shape; and, if turned over, they can easily right themselves.

EXPLANATION OF PLATE II. Skeleton of Starfish, *Asterias forbesi.* (Natural size.)

Photograph of dried skeleton from aboral surface, showing the reticulated skeletal plates bearing conspicuous spines. The skin has been mostly removed. The madreporic plate appears as a circular white spot between the bases of the two lower arms, while the circle of six small spines just to the left of the center of the disk indicates the position of the intestinal opening.

Their rapidity of growth is largely dependent upon the abundance of their food supply. Very young starfishes may double their size within a few days when food is abundant. The older ones grow comparatively more slowly, but far more rapidly than was formerly supposed. The rate of growth of our common starfish is discussed on page 40.

The previous food supply has a great effect on the size of the animal when sexual maturity is reached, some individuals being several times the size of others at the first breeding season. After the discharge of the sexual products growth is resumed, and may continue year after year, the oldest animals being as a rule the largest.

When no food is to be obtained, however, the starfish can live for months without apparent loss of vitality, the size being somewhat decreased in the meantime.

EXTERNAL STRUCTURE

The class Asteroidea, which comprises the starfishes, is distinguished from the other classes of Echinoderms by the presence of a central disk which merges gradually, without sharp demarcation, into the broad, radially arranged arms, or rays. The rays themselves are hollow, and contain portions of the digestive and reproductive systems and other organs.

The upper surface of disk and rays is termed the aboral, or abactinal surface, while the lower surface, having the mouth in the center of the disk, is the oral, or actinal, surface.

Between the bases of two of the rays on the aboral surface is situated a peculiar skeletal plate, the madreporic plate, or madreporite. In spite of its small size this plate is often conspicuous from the fact that it is of a different color from that of the adjacent surface of the body. If examined with a lens it may be seen to be pierced by numerous irregular openings, by means of which the sea water from the exterior may enter the water-vascular system (Plates II and III).

This madreporic plate may indicate the plane of symmetry in the body of the starfish, for a line passing through this plate and the axis of the ray opposite will divide the body into two nearly symmetrical portions, sometimes spoken of as right and

PLATE III. Aboral surface of Starfish, *Asterias forbesi.* Photograph of living animal. (Natural size.)

left. The two symmetrically placed rays adjacent to the madreporic plate constitute the bivium, while the three remaining rays form the trivium.

Body Walls. — The body walls, of both disk and arms, are supported by a network of articulating calcareous plates, or ossicles (Plates II and VIII), held together by connective tissue and muscular fibers. Imbedded in the body walls everywhere over the surface of the body are blunt calcareous spines, arranged in a more or less definite order, varying somewhat in different species, and often movable upon the underlying plates.

Covering the whole surface of the body, including even the spines and pedicellariæ, is a delicate membrane or skin, clothed externally with closely placed vibratile cilia (Plate III). The rapid vibration of these cilia keeps the sea water surrounding the body in constant motion.

Branchiæ. — In the smooth area between the spines the body wall is furnished with a multitude of protrusible filaments, the branchiæ (Plate I, fig. 4; Plate III), or respiratory processes, which have very thin walls and are filled with the fluid of the body cavity. When fully extended, an exchange of gases may occur between the contained cœlomic fluid and the external sea water, thus serving the purpose of respiration.

Pedicellariæ. — Scattered over the surface of the body in many species of starfishes, and arranged in groups at the bases of the spines (Plate III), occur great numbers of minute forceps-like or scissors-like appendages, called pedicellariæ. Each of these organs is provided with strong muscles which enable it to close tightly under the proper stimulus and thereby seize any tiny object with which it may come in contact. In the smaller pedicellariæ the two blades cross like a pair of scissors, while the others are like straight forceps (Fig. 8). By passing the objects from one to another, the pedicellariæ can by their united

---

EXPLANATION OF PLATE III. Aboral surface of starfish, *Asterias forbesi.*

Photograph of living animal. (Natural size.) Each of the large, blunt spines is surrounded by a circle of tiny whitish pedicellariæ, while other pedicellariæ and numerous minute tubular branchiæ are scattered irregularly over the dark colored skin between the spines. The madreporic plate is seen as a dark, circular area between the bases of the two lower arms.

efforts remove any small foreign body from the surface of the disk or arm and thus serve to keep the body clean. They also serve to protect the delicate skin from attacks of such small enemies as might otherwise cause serious injuries.

In the blood starfish (*Henricia*) pedicellariæ are absent, the body in this instance being protected by the multitude of closely set spines (Plate I, figs. 1—3; Plate XI), which cover all parts of the aboral surface.

Mouth.— Turning now to the oral surface of the body, the mouth may be seen to occupy the central portion of the disk. Surrounding the mouth are five groups of movable spines (oral spines, or mouth papillæ), forming the five angles of the mouth (Plate VIII). By means of muscular contractions the oral spines may be brought together to close the mouth opening, or may be opened widely to allow the extrusion of the stomach, as indicated in the two photographs on Plate VIII.

Tube-feet. — Extending outward from the mouth are five deep grooves, the ambulacral grooves, one of which occupies the oral surface of each arm (Plate IV). In these grooves are situated the locomotor organs of the starfish, consisting of a multitude of finger-like sucking tubes — the tube-feet — arranged in definite rows throughout the length of the arm. In the species of *Asterias* (Plates IV and IX) there are four such rows of tube-feet, while in *Henricia* (Plates I and XI) there are but two. The rows are placed symmetrically on each side of the axial line, as shown in Fig. 2.

Each tube-foot consists of a muscular tube with a circular disk at the free end (Fig. 1). The tube leads between the ossicles forming the roof of the ambulacral groove, and connects internally with a sac-like swelling — the ampulla, — and also, by a narrow tube, with the radial canal of the water-vascular system. The relations of these parts are shown in Fig. 1.

PLATE IV. Oral surface of starfish, *Asterias forbesi.* (Natural size.)

The ambulacral grooves are widely opened, disclosing the four rows of crowded tube-feet, each with its terminal sucking disk. In the center of the disk appears the delicate, baglike stomach, partially everted, as in the beginning of the act of feeding. The photograph shows two rows of slender adambulacral spines along the borders of the ambulacral grooves, and the large blunt spines, surrounded by minute pedicellariæ, nearer the sides of the arms.

PLATE IV. Oral surface of Starfish, *Asterias forbesi*. (Natural size.)

Both ampullæ and tube-feet have highly muscular walls, and both are filled with fluid from the water-vascular system. When, in locomotion, the ampulla of any particular tube-foot is contracted, the fluid contained therein is forced into the tube-foot itself, extending this organ until it becomes a long, finger-like process. On coming in contact with any external object the disk at the end of the tube-foot may be pressed tightly against it. If now the muscles of the tube-foot contract, the water which

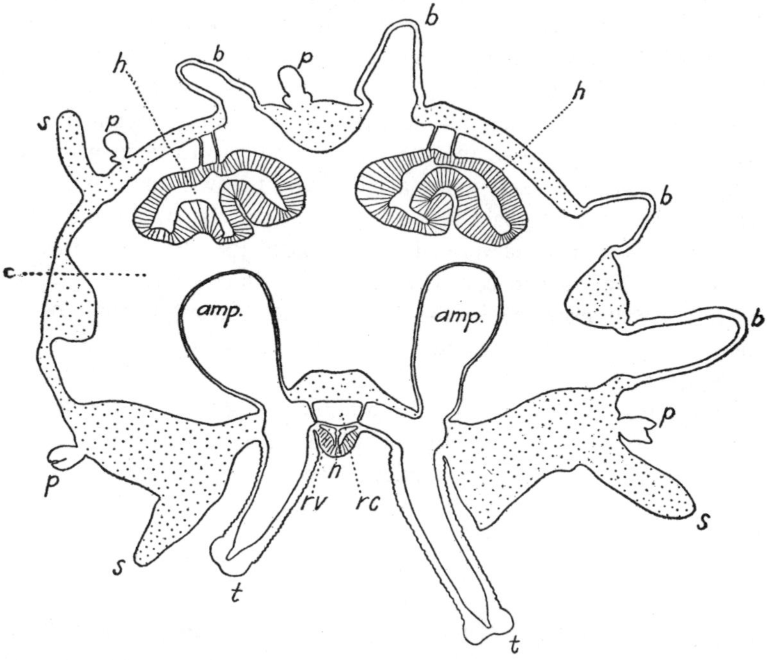

FIG. 1. Transverse section of a decalcified ray of a young starfish, showing the relations of the internal organs. The two ampullæ (*amp*) open by wide canals into the tube-feet (*t*), the water supply of which is provided by the radial canal (*rc*). The openings into the ampullæ from the radial canal are furnished with delicate valves, as indicated, which prevent the back flow of the water when the ampullæ are contracted and the tube-feet extended. Beneath the radial canal are the two radial perihæmal vessels (*rv*), separated by a vertical septum, and beneath these is the radial nerve-cord (*n*), which extends from the circum-oral nerve to the tip of the ray. Sections of the spines (*s*), branchiæ (*b*), and pedicellariæ (*p*) are indicated. The pyloric cæca (*h*) are attached by mesenteries to the dorsal wall of the ray. The spaces between the organs constitute the cœlom (*c*), and are in life filled with a fluid somewhat resembling the blood, and like it containing floating corpuscles.

it contains is returned to the ampulla, and at the same time the foot tends to shorten. As the terminal disk adheres to the external object by the pressure of the water, the shortening of the foot will tend to draw the body of the starfish towards the point of adhesion. If many such tube-feet work together, an actual movement of the body occurs. By the coördination of the nervous system of the animal all the tube-feet of any arm or of all the arms may be extended in the same direction, either to the right or left, backwards or forwards, as the case may be. The result is a slow, steady movement of the body.

Suitable muscular contractions of the other body muscles, combined with the working of the multitude of tube-feet, enable the animal to bring about a great variety of movements, by which it seeks its food, migrates from shallow to deep water or the reverse, rights itself when accidentally thrown upon its aboral surface, and forces apart the valves of the shells of the mollusks on which it feeds, as described on page 30. By such means, moreover, the starfish can so change the shape of its body as to occupy small crevices in the rocks, creep through small openings, and adapt itself to any surface with which it may come in contact.

Ossicles. — The roof of the ambulacral groove is formed by a series of interlocking movable plates, the ambulacral ossicles, arranged in a double row articulating with one another at the apex of the groove in such a way as to open or close the groove when required (Plate VIII).

On each border of the ambulacral grooves is a regularly arranged row of plates, adambulacral ossicles, which bear the adambulacral spines (Plate VIII). These spines may all be turned toward the middle line of the groove in such a way as to form a protection for the tube-feet.

The two rows of ambulacral ossicles end in a single terminal ossicle at the tip of the arm. This ossicle supports a terminal tentacle, endowed with a sensory function, presumably both tactile and olfactory in nature. At the base of the tentacle the radial nerve ends in a minute, orange-colored, eye-spot.

Between the ambulacral ossicles are the pores by which each of the tube-feet is connected with the corresponding ampulla, situated internal to the ossicle. The firmly interlocking ossicles

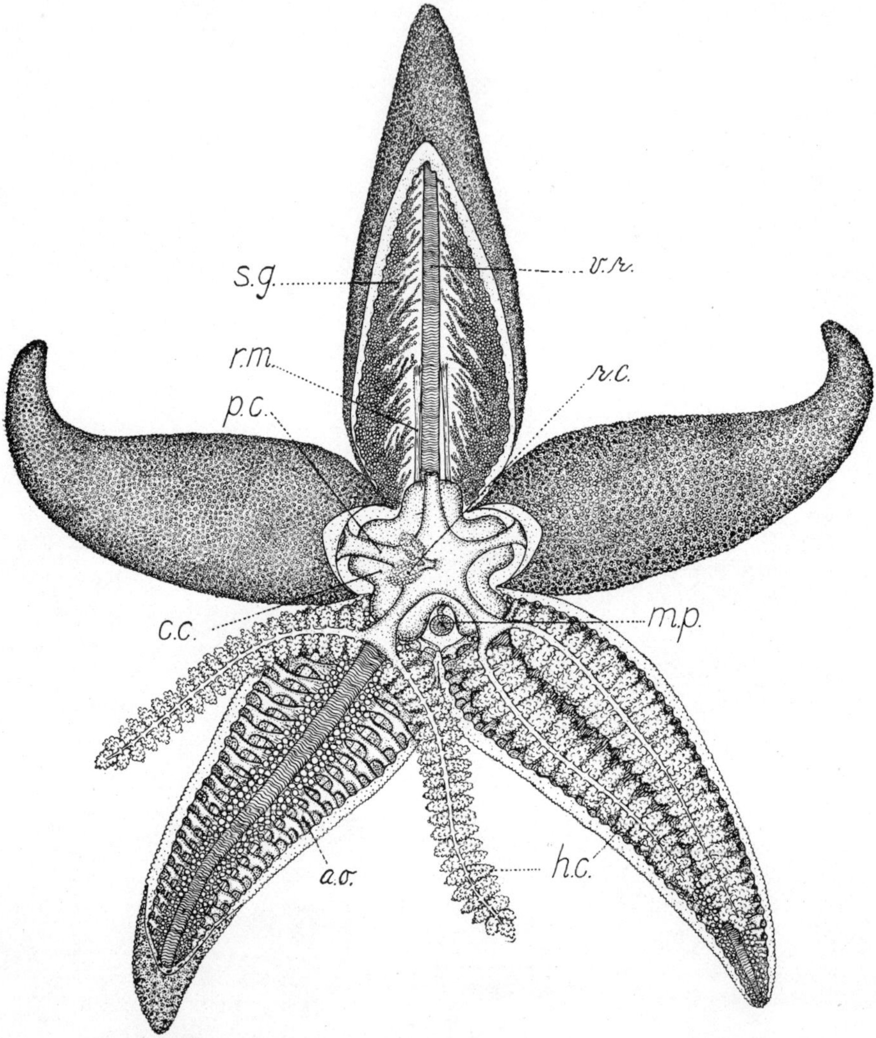

PLATE V. Anatomy of Starfish, *Asterias vulgaris*. (Two-thirds natural size.)

which form the skeleton of aboral and lateral surfaces of the disk and arms are shown in Plate II.

INTERNAL STRUCTURE

If the body of a starfish be opened by removing with fine scissors the body walls of the aboral surface of the disk and one or more arms, as indicated in Plate V, the internal organs are easily exposed. The most conspicuous viscera are the various portions of the digestive system, and, in certain seasons of the year, the reproductive organs.

In the dissection shown on Plate V, the five cardiac pouches of the stomach (*c. c*) lie beneath the five-lobed pylorus (*p. c*), into which open the profusely branched digestive glands, pyloric, or hepatic, cæca (*h. c*). The madreporic plate (*m. p*), rectal cæca (*r. c*), ambulacral braces (*a. o*), retractor muscles of stomach (*r. m*), vertebral ridge (*v. r*) and sexual glands (*s. g*) are shown. In the lower left-hand ray the pyloric cæca are laid aside and the sexual glands removed to show the ampullæ of the tube-feet on each side of the vertebral ridge. In the median ray the hepatic cæca are removed completely to show the position and extent of the sexual glands (*s. g*).

Alimentary Canal. — The digestive system of the starfish is modified to conform with the general shape of the body. The small, centrally placed mouth leads by a short esophagus into the spacious cardiac portion of the stomach. The latter is provided with five broad lobes, cardiac pouches, opposite the bases of the five arms (Plate V). The walls are very thin, much folded, and muscular, so that this portion of the stomach can be completely everted from the mouth opening and wrapped around such food as the animal secures. After the prey has been

---

Explanation of Plate V. Anatomy of starfish, *Asterias vulgaris*. (Two-thirds natural size.)

Dissection from aboral surface to show internal anatomy; *a. o*, ambulacral braces; *c. c*, cardiac portion of stomach; *h. c*, hepatic cæca; *m. p*, madreporic plate; *p. c*, pylorus; *r. c*, rectal cæca; *r. m*, retractor muscle; *s. g*, sexual glands; *v. r*, vertebral ridge.

The lower right ray shows the viscera in their natural position. In the lower left ray the sexual glands and retractor muscles have been removed and the hepatic cæca separated to show the ambulacral ossicles and ampullæ of tube-feet. In the upper ray the hepatic cæca have been removed.

completely digested, the stomach is again drawn back into the body cavity by means of five pairs of long retractor muscles, which are attached to the upper side of the ambulacral ossicles in the rays, as shown in Plate V.

At its aboral end the cardiac stomach is constricted and leads into a narrower cavity, the pylorus, from which five pairs of large, profusely branched pyloric, or hepatic, cæca (Plate V) extend outward into the cavities of the rays. These cæca form the digestive fluids, which are able to act on fats, proteids, and carbohydrates. From the pylorus a short and narrow intestine (Plate V) leads to a minute anal opening on the aboral surface near the center of the disk, and on a line between the bases of the left and middle rays of the trivium. Connected with the intestine are two small glands, the rectal cæca (Plate V), which doubtless excrete waste products, to be discharged from the body through the anus.

As stated on page 32, whenever any considerable amount of refuse remains after digestion such substances are thrown out of the mouth. Consequently the intestine has little use except as an excretory organ.

Water-vascular System. — The complex system of vessels constituting the water-vascular, or ambulacral, system are of great importance in the locomotion of the body, as described above under the heading, "Tube-feet."

The madreporic plate serves as a sieve (*m*, Fig. 2) through which the external sea water enters the water-vascular system. This plate connects with the madreporic canal (the so-called stone canal), which leads into a large circular vessel surrounding the mouth. From this vessel a large radial vessel runs along the roof of the ambulacral groove to the tip of each arm. The radial vessel gives off paired lateral branches to the tube-feet (Fig. 2). The latter are all arranged in two rows on each side, and those of the two rows alternate. Consequently longer and shorter lateral branches alternate. The action of these tube-feet in locomotion is explained on page 23. Connected with the internal border of the circular vessel is a series of nine glandular sacs known as Tiedemann's bodies, shown in the diagram (Fig. 2). These sacs give rise to amœbocytes, or cells which float about in the fluid of the water-vascular system. They probably serve an important function in the collection and discharge of the waste products of the body.

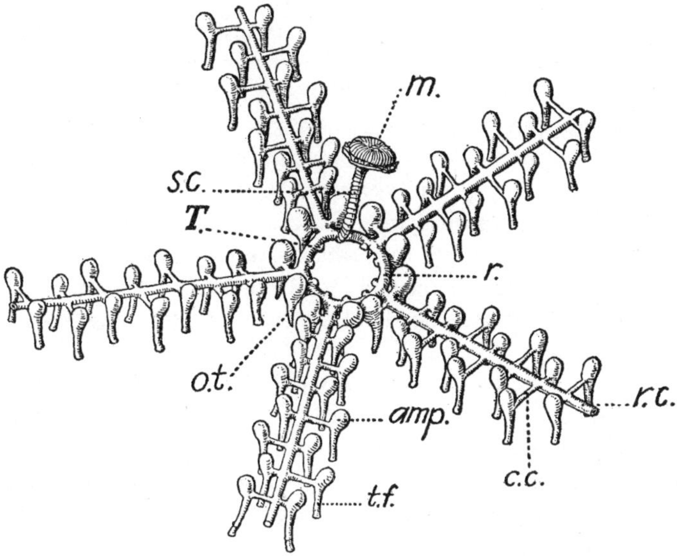

FIG. 2. Diagram of water-vascular system of starfish; *m*, madreporic plate; *s. c*, stone canal; *T*, Tiedemann's body; *o. t*, oral tentacle; *t. f*, tube-foot; *amp*, ampulla; *r. c*, radical canal; *c. c*, connecting canal; *r*, ring canal.

Circulatory System. — A complex system of vessels running parallel with the vessels of the water-vascular system is termed the perihæmal system. The principal vessels of this system are divided into two parallel chambers by a vertical septum. In the septum is a smaller central canal which contains a small amount of fluid, and is thought to constitute the true blood system. This latter has therefore radial and circular canals, parallel both to the water-vascular and to the perihæmal system. The radial vessels are shown in section in Fig. 1.

Nervous System. — The nervous system consists of a nerve ring surrounding the mouth, with a large radial nerve extending outward along the roof of the ambulacral groove of each arm and terminating in an orange-colored eye-spot at the tip of the arm. From the radial nerves minute branches are given off to each tube-foot and ampulla, to the superficial integument, and to the muscles and organs in the body walls. In addition to these main nerves is a complex network of fibers supplying the alimentary canal and other internal organs. Everywhere over the surface of the body is a network of minute fibers connected with the sense organs, pedicellariæ, and branchiæ of the integument.

The whole nervous system controls the coördinated actions of all parts of the body only when the connection with the nerve ring is intact. If the radial nerve at the base of one or more of the arms is cut, each arm so isolated from the nerve ring will act independently of all the others, and no coördinated movements of the body are possible. The nerve ring may therefore be looked upon as the coördinating center of the nervous system of this animal.

Reproductive System.— The sexes are separate, with paired sexual glands extending into each arm. These glands develop in close connection with the madreporic canal in a mass of tissue known as the axial organ. The development of the sexual products is described in detail in the chapter on reproduction.

HABITS

Many interesting papers have been published in recent years describing in great detail the habits of various species of starfish and their responses to stimuli. In addition to the classical works of Romanes* and Preyer,† the papers of Cole,§ Cowles,¶ Jennings,** von Uexküll†† and others discuss the behavior of these animals when subjected to a great variety of conditions.

* In addition to several more technical papers, the book entitled, "Jellyfish, Starfish, and Sea-urchins," published in the International Scientific Series, 1885, contains a most interesting popular account of the natural history of the starfish.

† Ueber die Bewegungen der Seesterne. *Mitteilungen aus der Zool. Station zu Neapel,* Band 7, 1887.

§ Cole, L. J., Direction of Locomotion of the Starfish (*Asterias forbesii*). Published in *Science,* vol. xxxi, 1910. This paper consists of a brief statistical study of the direction of movement of our common starfish in the absence of directive stimuli.

¶ Cowles, R. P., Reaction to Light and other Points in the Behavior of the Starfish. In Publication No. 132, Carnegie Institution, Washington, 1909. Describes the reactions of two species of starfish when subjected to various kinds of light.

** Jennings, H. S., Behavior of the Starfish *Asterias forreri* de Loriol. In the University of California Publications in Zoölogy, vol. iv, 1907. Observations on the habits of a species of Pacific Coast starfish and its reactions to various kinds of stimuli.

†† In addition to several technical papers by von Uexküll, a popular account of the general behavior of the starfish is contained in his recent book entitled, "Umwelt und Innenwelt der Tiere."

The general mode of life of the starfish, including habitat, locomotion, changes in shape of body, and migration, are briefly discussed in the following chapters, and the nature of this Report does not justify further details. The subject of habit formation, however, as described by Jennings* may be briefly outlined here.

"The behavior of the starfish is in general characterized by what might be called versatility; the animal can do in many different ways anything that it can do at all." In its locomotion, for example, it may move with equal facility in any direction along the plane of its body without any regard to right or left, or forwards or backwards. Under the same conditions its behavior is subject to the greatest variations, and it may be said never to do the same thing twice in exactly the same way.

When subjected to experiment it was found to solve the problem set for it, such as escaping from cramped quarters, removing rubber bands from its rays, etc., in a great number of different ways on successive occasions. But whenever left to itself it showed no improvement in solving these problems. It never learned to select the most effective method; that is, it failed to form the habit of employing this method when confronted again with the same problem.

But attempts to train the starfish were not altogether unsuccessful; for example, when placed on its aboral surface, the animal may right itself by employing any two adjacent arms, different pairs being used on successive occasions. Yet by preventing the use of certain arms the animal may be successfully trained to use certain other arms on all occasions. But the effects of this training are soon lost, so that on the following day the starfish shows no effect of its training. By giving the animal ten lessons a day for eighteen days, however, the starfish may show the effect for at least a week. By long continued training it is probable that still more lasting habits might be induced.

FOOD

One of the species of asteroids occurring in Connecticut waters, the common starfish (*Asterias forbesi*), causes a vast pecuniary loss each year because of the enormous numbers of

* Johns Hopkins Univ. Circulars, No. 3, 1907; also Univ. Calif. Publications in Zoölogy, vol. 4, 1907.

oysters which it devours. The extent of this loss is difficult to estimate, and the reader is referred to page 33 for an account of the injuries thus caused.

The most usual food for all the species consists of mollusks and crustaceans, of which the various species of mussels, the oyster, the two common species of clams, and barnacles, furnish the principal supply. Univalves, or sea snails, of various species, and other invertebrates, are devoured whenever they can be secured; and indeed almost any small animal when injured or dead may be eaten. The larger starfishes may even devour small individuals of the same or of different species. The common starfish is occasionally of actual benefit in that it devours the univalve known as the oyster drill (*Urosalpinx cinerea*), which is itself a serious enemy of the oyster. But it does this only when oysters or other bivalves are not obtainable.

The starfish secures its prey by seizing it with the tube-feet and then drawing it toward the mouth. The mouth, however, is itself of small size as compared with the body of the starfish, being only about one-fourth to three-eighths of an inch in diameter in a moderately large animal. Only objects of small size can therefore be taken directly into the stomach. If the prey be too large to pass the mouth, which is the case with the greater part of the food, the soft, sac-like stomach is everted from the mouth and wrapped around the prey, after which digestive fluids are poured out and digestion ensues. After the absorption of the nutritive substances contained in the body of the prey, the stomach is again drawn into the body of the starfish, by means of a special set of retractor muscles.

If a large bivalve be secured, as a large oyster, clam, or mussel, the shell must be opened before the food can be devoured. But it is obvious that the strength exerted by the tube-feet of the starfish will have little immediate effect on the powerful muscles which hold together, and close tightly, the two valves of a large oyster or quahog. Many theories have been proposed in the attempt to explain the process by which these mollusks are opened by the starfish. It was formerly thought that the starfish excreted a poison which found its way between the valves of the shell and finally stupefied the mollusk sufficiently for the valves to gape and allow the starfish to insert its stomach.

But experiments show that a mollusk taken away from a starfish which had opened it, soon regains its normal condition. Another theory was that the starfish wrapped its stomach about the mollusk's shell, depriving the animal of oxygen and effectually smothering it. But observation shows that such is not the case. Another theory that has been held is that an acid dissolves the edges of the shell, but no such acid is present. Others have believed that the edges of the shell are chipped sufficiently to allow the entrance of the starfish's stomach; and it is true that the delicate edges are often accidentally broken by the starfish, but such a shell can still be closed so tightly as to hold water.

Thus the problem remained unsolved until 1896, when the experiments of Schiemenz* showed that simply the long continued pull of the starfish's arms will eventually overcome the resistance of the adductor muscle which holds the two valves of the shell together. While the muscle can resist for a short time a very strong pull tending to separate the valves, the muscle eventually becomes fatigued, and relaxation allows the valves to gape. In the same way, if a sufficiently stout cord is tied around the foot of a large marine snail, a strong man is unable to pull it from the shell. Yet such a snail is unable to support its own weight, by the same muscles, for a long period. After being suspended for a time by the foot, the muscles become fatigued and relax, allowing the body and shell to hang loosely at the end of the fully extended foot.

Schiemenz demonstrated that the starfish was able to exert a pull of more than 1,300 grams, while it required a force of only 900 grams to fatigue the muscles of a large bivalve so that the shell would gape widely within a half-hour.

It is only necessary therefore for the starfish to grasp the oyster with the tube-feet of opposite arms attached to the valves, and then continue pulling in opposite directions until the mollusk's adductor muscle becomes sufficiently fatigued to allow the valve to open. But a considerable force is required in order to accomplish this, and it is only when the arms of the starfish are wrapped around the shell of the bivalve in such a manner that the tube-feet in the basal portions of the arms can

* *Mitteilungen des deutschen Seefischervereins*, 1896.

pull at right angles to the valves that the shell can be opened. Thus a starfish finds it quite impossible to open a bivalve when held with its arms outstretched beween two glass plates, even when there is plenty of room for it to carry the bivalve about with it.

After the resistance of the adductor muscle of the mollusk is completely overcome, the stomach of the starfish is turned out between the two valves of the shell, and the soft body of the mollusk is then digested as in the case of a smaller object, leaving nothing behind except the cleaned and empty shell. Sometimes two or more starfish participate in devouring a single large oyster.

The processes concerned in the digestion of a large mollusk are so complete in the starfish that little if any refuse remains. But in the case of animals sufficiently small to be taken directly into the stomach, through the mouth, the indigestible food remains are thrown out of the mouth again. For this reason the opening of the intestine on the aboral surface of the body is very small, and serves merely for the discharge of fluid waste materials secreted by the glands connected with it.

When food is abundant the starfish often devours an enormous quantity, and growth is correspondingly rapid. The observations of Mead, on the voraciousness of young starfish, as noted on page 57, show that as many as 56 clams, some of which were as long as an arm of the starfish, could be devoured by a single starfish within a period of six days.

On the other hand, in case food cannot be secured, the starfish is able to survive a fast of several months without apparent discomfort. In the latter case, however, no growth whatever occurs until food is again secured.

NATURAL ENEMIES

It would seem that an animal so well protected by its spiny skin and abundant pedicellariæ, would be largely exempt from the attacks of other carnivorous animals of the sea. And in the adult condition such is actually the case. Occasionally one starfish when pressed with hunger will devour another of the same species; fishes of various kinds destroy mutilated individuals and probably large numbers of young. Various species of birds, as crows and gulls, which frequent the shore, sometimes feed upon the starfish when other food is difficult to obtain. The starfish is said also to suffer occasionally from the effects of a parasitic disease.

But the greatest destruction occurs in the very early stages in the life of the starfish, at the time when the bipinnariæ or brachiolariæ, as the embryos are termed, are swimming free at the surface of the water. At this time such fishes as the menhaden swim through the water with their mouths widely open and devour countless numbers of small organisms of all sorts from near the surface. Among these may be the young of two species of *Asterias,* and vast numbers are destroyed in this way. The young of *Henricia* develop directly from the egg, and thus escape the dangers of the free-swimming stage.

How does it happen, then, if there be so few natural enemies of the starfish after it has assumed the adult condition, that the sea is not overrun with the creatures to the exclusion of all other animals? Of the thousands of eggs laid by a female starfish each year not more than one on the average produces a sexually mature animal. All the others die; some are destroyed by the enemies indicated, many die from their inability to obtain food, others are buried beneath the mud of the bottom, some are crushed by the rolling pebbles on the shore, still others succumb to changes in the temperature of the water, many die from the effects of rain or streams of fresh water, while vast numbers are left by the tides to be destroyed by drying in the air.

As a result of these many hazards only a small fraction of one per cent. of the embryos survive to produce young starfishes a half-inch in diameter. The few that reach this stage of development are comparatively safe.

DAMAGE TO OYSTERS AND OTHER MOLLUSKS

There is little realization except among the oyster planters themselves of the enormous extent to which the starfish is destructive to the oysters and clams which form such an important natural resource of the state. The oyster industry alone produces an annual income of over a million dollars, and furnishes employment to hundreds of citizens.

For the year ending June 30, 1910, the receipts in this state from the sale of market and seed oysters and shells was $1,892,759, while there remained on the beds in private areas alone an estimated quantity of 3,784,175 bushels of oysters, valued, with the stakes and buoys which mark the grounds, at $1,856,285. During this year the industry gave employment to 2,703 persons.*

For the mere privilege of controlling certain parts of the floor of the Sound for the cultivation of oysters, the oyster planters have paid over $260,000 into the treasuries of the state and the towns adjoining the coast.†

So serious is the starfish injury to the oyster industry that the General Assembly of this State, in 1901, passed a law making it illegal for any person to assist in the spread of the pest. The law reads as follows: "Every person who shall wilfully deposit or assist in depositing any starfish . . . . in any of the navigable waters of this State . . . . shall be fined not more than fifty dollars, or imprisoned not more than six months." (Connecticut Shell-fish Commissioners' Report for 1903, page xxvi.)

The actual loss which the starfish causes to the oyster industry of the state is very difficult to determine with any degree of precision. Any estimate even of the losses to a single grower is, from the very nature of the case, largely a matter of guess-work. Furthermore the natural beds suffer a great annual loss, of which no one takes account. The loss to the new set, when the individual is but a small fraction of an inch in diameter, may actually be, and probably often is, greater than that caused by the more conspicuous destruction of oysters of larger size. The oyster grower who watches his plantations carefully knows when the starfishes of larger size become numerous on his beds, but he has no means of determining the injuries caused by the young individuals to correspondingly young oysters. The fact that a single starfish less than a month old has been shown to be able to devour more than 50 young clams in six days, indicates how extensive these injuries may become.

---

* Report to the General Assembly, State of Connecticut, on the Investigation of Oyster Properties by the Board of Equalization, 1910, page 77.

† Report Shell-fish Commissioners, 1907-08, page 82.

On the privately controlled beds regular examinations are made for the presence of the starfishes, and, whenever they are found in unusual abundance, it is imperative that immediate steps be taken to destroy them. This is accomplished by the use of mops, as described below, by means of which the larger starfishes are caught and killed.

By such thorough and regular inspection the very extensive injuries of former years have been largely controlled, and the oyster grower of to-day, like the fruit grower on the land, always reckons upon a certain expense for the destruction of pests.

The actual expense of this method of controlling the starfish varies somewhat from year to year; and, while an accurate estimate of the total amount is not available, it is known to be very great. And yet, as in the case of the fruit grower, it is but a fraction of the amount that would inevitably be lost if such measures were not taken.

During the year 1909, inquiry as to the number of starfish actually destroyed in carrying out this policy of protection shows that many thousand bushels of the pests are annually destroyed. When we consider the fact that each bushel will contain several hundred starfish, the number, of course, varying with the size, we may gain some conception of the vast number destroyed.

In reply to an inquiry, the following statements by a prominent oyster planter may be considered as expressing the most accurate information obtainable. "The starfish seem to vary in different localities from month to month, but the average on the whole seems to keep up remarkably high, considering the fact that such a vast quantity of these destructive creatures are caught by the oyster steamers." "We have not at command definite amounts that are caught per year, but in a general way would say that they probably run from 5,000 to 10,000 bushels per year. Some years they may exceed the latter figure."

The excessive abundance in restricted localities often occasions great alarm among the growers, but of late years success has invariably rewarded the determined effort to combat the pest upon any particular bed. Sometimes, however, it must be admitted that this result has been attended with very great expense.

Although it is probable that the injuries during recent years are on the whole less than in some former years, yet the total loss is still very great. In December, 1909, for example, the following item from a Rhode Island paper indicates that even during that year the damage in certain localities must have been excessive.

"The fishermen and oyster growers are complaining of the prevalence of starfish, which are doing great damage to the growing stock. Old oystermen say that never in the history of the business have the starfish been so plentiful, and many of the steamers in Cowesett bay and other points along the coast, which in ordinary seasons are engaged in dredging, are now dragging 'starfish mops' over the beds to get rid of the pest.

"One prominent oyster man says that the damage to the oyster beds in Rhode Island and Connecticut this season will be at least a quarter of a million dollars, when the loss of stock and the expense of catching and killing is taken in consideration. The growers are at a loss to explain the unprecedented increase of the pest other than last summer the weather conditions were right for breeding the fish. The expense of catching the starfish is practically so much money lost, as the stars are of little use except as fertilizer, and even for that purpose they do not command much in the way of price, as they are not caught in large quantities."

The point of view of the oyster planter is well expressed in the following statements by Mr. B. Frank Wood, Superintendent of Shell-fisheries of New York State.*

"In Long Island Sound, the starfish is the enemy most to be dreaded. The set of stars seems to have some relation to the oyster set, and, in seasons when the young oysters are abundant, the planters look for a like abundance of stars.

"This fish is certainly an ill and evil star to the oyster and to its cultivators. At times covering the bottom to a depth of eighteen or twenty inches and extending in solid bunches or masses over considerable areas of ground, they are capable of blanketing entire beds of oysters in their slow but sure advance, leaving not a living bivalve behind. They are very hardy and tenacious of life, and may even be deprived of some of their limbs or rays and regain these members by a new growth which makes good the loss. Unless these pests are constantly attacked

* Biennial Report of the Shell-fish Commissioners of Connecticut, 1907-08, pp. 96-99.

in season and out of season, and their numbers reduced, they are sure to overwhelm and totally destroy the oyster beds. This refers particularly to shell-fish lands situated in Long Island Sound, for it is true that there are some localities where comparative freedom from the enemy is enjoyed.

"Many devices have been invented and used for the purpose of destroying the stars. The one now in general use by the planters is the star-mop, or tangle. It is made of cotton cords or strings, arranged in large tassels or bunches, attached to a steel frame, and drawn over the beds by means of the dredging chains and machinery. The stars become entangled in the meshes of these mops and are raised in large numbers. Mr. Herman D. Pausch has made many experiments for the purpose of perfecting some better plan of combatting this scourge. He has found, he believes, a practicable method for use in localities where there is not too great depth of water. His plan is to make a continuous wall or ridge of lime along the boundary of the bed to be protected. He has accomplished this by filling paper bags with quicklime and dropping these bags along the line. The paper serves to hold the lime from being carried away by tides while descending through the water. The water will of course slack the lime, but Mr. Pausch states that so long as the lime barrier remains intact no starfish will cross it. He has experimented quite extensively in this direction, and is of the opinion that the results obtained will warrant the use of his methods upon a large scale. One of his tests consists of placing starfish within lime enclosures, and though kept there for considerable periods of time not one will attempt to cross the line which separates it from freedom.

"Mr. Pausch says that the lime barrier constitutes a veritable dead line for the sea-star. He considers the use of paper bags a somewhat clumsy plan for getting the lime to the bottom, and is now engaged in perfecting an apparatus by means of which he may feed the lime through a hose or pipe so drawn over the ground as to leave an unbroken ridge of the material on the bottom.

"The vast numbers with which the star overwhelms the shell-fish beds, taken in connection with his insatiable appetite, make it clear that nothing but eternal vigilance on the part of the planter will hold the day against him.

"There is, consequently, a community of interest among the planters, in attacking and destroying the star. For, in protecting his own beds, a man also assists his neighbor to whose tracts the stars might go, and, conversely, any neglect of one's own ground constitutes a menace to all cultivators in the vicinity.

"The work of destroying stars is a duty, as before intimated, in which the natural-growth oysterman usually takes no part. He takes the oyster where he can find it, appropriating the bounty of heaven without accepting any responsibility as to the conservation of a further supply."

Whenever the starfishes are found upon an oyster bed in excessive numbers, the oyster grower sends out his boats with the star-mops, or tangles, which are dragged back and forth systematically over the bed.

"When the star-mop is drawn to the surface, its load of stars is deposited in a vat or tank and treated with boiling water or live steam. The oyster boats are well equipped with apparatus for this purpose, the steam or water being conveniently taken from the vessel's boiler, though some of the steamers have special boilers for this purpose.

"Attempts have been made to turn to some useful purpose the large quantities of stars taken from the oyster beds of Long Island Sound, and it goes without saying that this should be a more or less valuable by-product of the industry. Properly mixed with other material, they constitute a valuable fertilizer. Frequently, farmers residing in the neighborhood of oyster planters have contracted for the catch of stars, but the oystermen say that in practice it does not work well, as they find that they cannot depend upon having the accumulation of dead stars regularly removed from their premises, and of course they cannot allow them to remain and putrefy. As a consequence, the usual mode of disposition is to throw them overboard after they have been killed in the live steam or boiling water bath."

A perusal of the Reports of the State Shell-fish Commissioners reveals many instances of the excessive losses caused by starfish in former years. In the report for 1883 the statement occurs that thousands of bushels of oysters were destroyed in one locality within a single week. A single firm lost $20,000 worth in one bed, while another firm suffered a loss of $100,000 in "the last two years."

The report for 1884 states that 11,000 bushels of the pests were caught off Stratford during a period of six weeks, but before the work was completed at least 200,000 bushels of oysters had been destroyed.

In a paper by Collins on the Oyster Fisheries of Connecticut, an estimate is made that in 1888 the damage to beds in Connecticut waters amounted to some $651,500. During that year 42,000 bushels of starfish were destroyed.

Appreciating the ability of the starfish to migrate from one spot to another not far removed, one enterprising firm of oyster growers has sometimes paid a dollar for each bushel of young starfish collected on the shores adjacent to their plantations. Such a procedure is doubtless profitable to all the parties concerned; for, when the young starfish are abundant, the dollar is easily earned, and the destruction of so large a number of small starfish as may be contained in a bushel basket — occasionally as many as 2,000 very small ones — will undoubtedly reduce the expense of protecting the beds by more than the sum expended.

Not only is there great destruction of oysters, but other mollusks of economic importance, as the clams and mussels, suffer, perhaps, to an even greater proportional extent. The mussel is one of the chief natural food supplies of the starfish, but, as this mollusk is little used commercially in Connecticut, the pecuniary loss is not of importance. The injury to the clam, on the other hand, which furnishes an important article of food, is a matter of some real concern to the people of the state. The barnacles, worms, crustacea, and various gastropod mullusks, from which the starfish derives a portion of its food supply, have little to do with human welfare; and, whenever the starfish supplements its diet by devouring one of its own kind, or when it feeds upon the mollusk known as the oyster drill, it becomes, for the time being, of distinct service. Likewise, in furnishing a portion of the food supply, either directly or indirectly, of certain fishes used for human food, it compensates to some slight extent for the damage which it causes.

RATE OF GROWTH

The rapidity with which starfishes increase in size, and the age at which sexual maturity is attained, have long been subjects for discussion among zoölogists. A number of years ago Alexander Agassiz estimated the growth and age of the common starfish, and came to the conclusion that it required about fourteen years for the animal to reach its full development. It was thought to begin to spawn before that time, apparently at six or seven years of age.

Professor Mead, however, has settled this point beyond dispute* by actually rearing large numbers of young starfishes from the time when the free-swimming larvæ attach themselves to the eelgrass and other objects in preparation for their transformation to the adult form.

By keeping these young starfishes in aquaria and submerged cars and feeding them regularly, a most surprising rapidity of growth was found to result. On June 29th the stars measured on the average only about 1mm. from mouth to tip of arm; on July 18th, 5mm.; July 26th, 9mm.; August 18th, 18mm.; Sept. 26th, 35mm.; and on Oct. 25th, 54mm. Such stages of growth are represented in their natural size on Plate VI, from Mead's report in the Bulletin of the U. S. Fish Commission for 1899.

Figs. 1 to 3, on Plate VI, show small individuals taken from seaweed soon after having completed the free-swimming stage. Fig. 4 shows the average size found on eelgrass and seaweed about two weeks later, July 15, and Fig. 5 a large individual of the same age which was reared in a submerged car and well supplied with food. A similar individual reared in car, three days later, is shown in Fig. 6; and one after eight days more, July 26, in Fig. 7. Figs. 8, 9, and 10 indicate the size of the young starfishes about four weeks, six weeks, and nine weeks after having completed the free-swimming stage. Fig. 11 shows the growth in three weeks more, while Fig. 12 indicates the size attained by a large individual reared in submerged car with an abundance of food, at about 4 months of age.

---

* Mead, A. D., On the Correlation between Growth and Food Supply in Starfish. *American Naturalist*, vol. xxxiv, 1900.

PLATE VI. Growth of Starfish, *Asterias forbesi*, during the first four months of life. (Natural size.) After Mead.

There were larger specimens than those indicated, and also smaller, for at all times there were some individuals which were only a third as large as the others, and all intermediate stages occurred between the smallest and the largest. Similar series could be obtained along the seashore at any time during the summer and fall.

By varying the amount of food supplied, Professor Mead was able to prove that the rate of growth depended directly upon the amount of food available. In this respect the starfish behaves quite differently from the higher vertebrate animals, for the starfish can assimilate many times as much food as is necessary for keeping it in good condition, and all the food not required for the maintenance of the bodily functions is used in the manufacture of new tissues, whereby a rapid growth results. Yet a starfish will live and remain in apparent health for months with scarcely any food whatever, although no growth takes place under such circumstances.

When the usual food supply becomes scarce, the larger starfishes show no hesitation in devouring smaller individuals of their own kind.

In Fig. 3 are shown outlines of two individuals of the same age, five and one-half weeks, the small star on the left representing a starfish to which but little food was supplied, while the larger figure shows the size which a well nourished specimen may reach at the same age.

EXPLANATION OF PLATE VI. Rate of growth of starfish, *Asterias forbesi.*

A series of young starfish to show the rate of growth of well-nourished individuals during the first four months of life. All natural size.

Figs. 1 to 3. Small individuals during the first week after having completed the free-swimming stage, July 1.
Fig. 4. Specimen from eelgrass two weeks later, July 15.
Fig. 5. Specimen reared in submerged car about three weeks after the completion of the free-swimming stage, July 15.
Fig. 6. Similar specimen three days later, July 18.
Fig. 7. Large specimen from car when one month old, July 26.
Fig. 8. Similar specimen on August 2.
Fig. 9. Similar specimen on August 18.
Fig. 10. Same on September 5.
Fig. 11. Same when three months old, September 26.
Fig. 12. Largest specimen reared in submerged car at the age of four months, October 25.

(After Mead.)

It is of interest to note that the loss of an arm does not cause even a temporary diminution in the rate of growth, the rest of the body increasing in size as rapidly as in the case of uninjured individuals. Even the loss of two arms has little effect. The lost arms are rapidly replaced if food is abundant, and in their regeneration grow faster than the others until the normal symmetry of the body is restored.

FIG. 3. Outlines of two starfishes of the same age (5½ weeks), showing difference in the rate of growth under the influence of different food supplies. (After Mead.)

The age at which the starfish becomes sexually mature is also influenced by the rate of growth, for Mead has found that all starfishes with a length of ray of more than 50 mm. at the beginning of the breeding season, in May, are sexually mature, while others, but 10 or 15 mm. long at the same date, do not deposit sexual products until the following year. Now the length of 50 mm. is reached by well nourished individuals when less than four months of age, so that it seems reasonable to conclude that the well fed starfishes are ready to breed when a little less than one year old; that is, in June of their first year.

The rate of growth is most rapid during the warmer months of the year, and may cease entirely during the winter. This is probably due in part to the smaller amount of food devoured, and in part to the development of the sexual products.

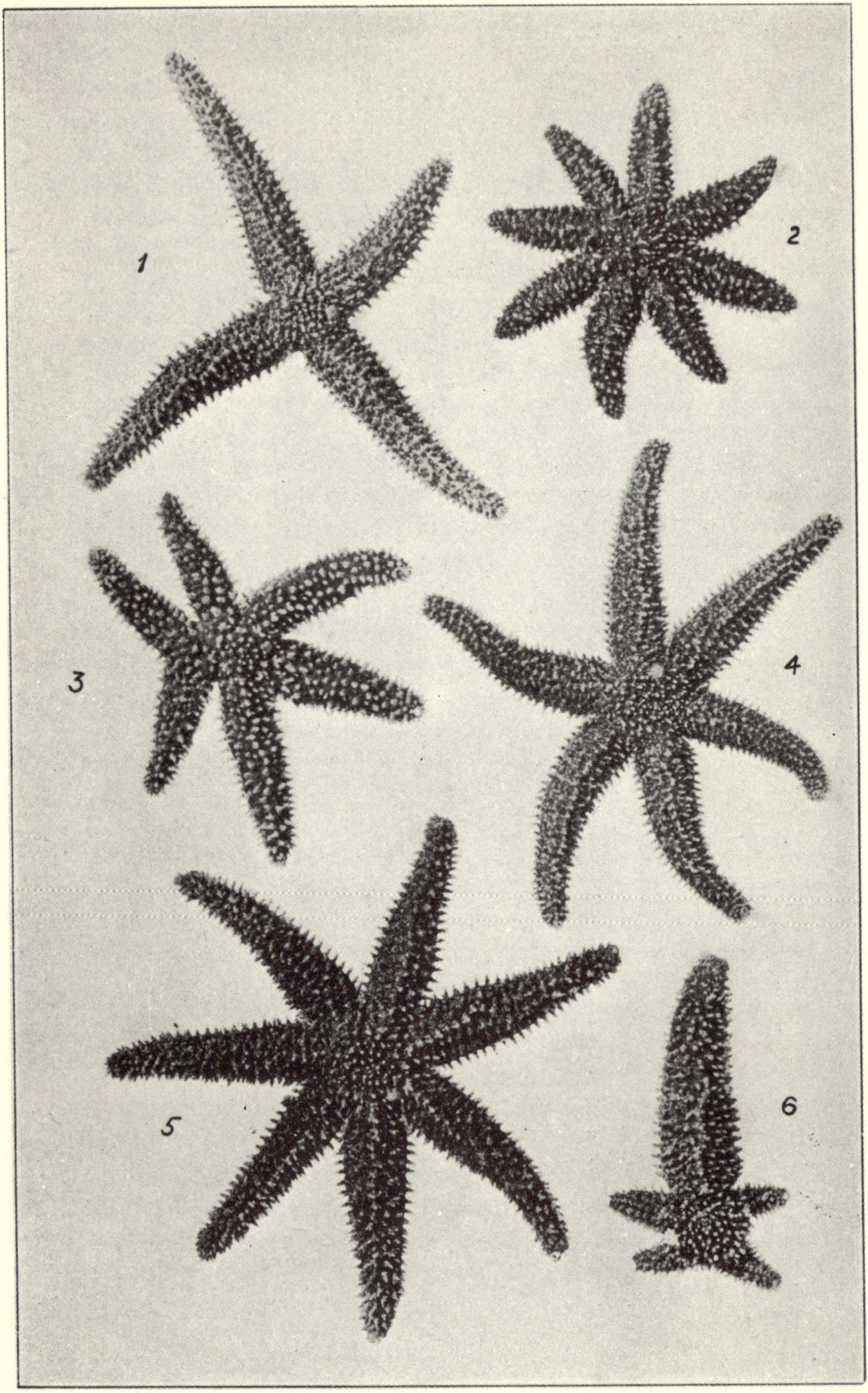

PLATE VII. Variation, *Asterias forbesi.*

MIGRATION

It is well known that when the food supply in one region is exhausted the starfishes migrate to other localities, and furthermore there is more or less regularly a seasonal migration from the tide pools along the shore to the deeper water. But the extent of such migrations has probably been greatly overestimated, and the sudden appearance of vast hordes in any particular spot is due more to the very rapid increase in the size of the small and previously unnoticed starfishes than to the actual migration of a multitude of individuals. If a starfish should move constantly in one direction at the rate of six inches per minute, which is about its normal rate of progress, it could travel about a mile per week; but there is no evidence that such a rate is maintained for more than a few minutes at a time, and it is probable that the creature is quite as likely to retrace its steps or to proceed in an irregular path as to continue for more than a few minutes in any single direction. As stated in the quotation on page 37, quicklime has proved an efficient barrier, although the animals can easily pass all kinds of soft or slippery substances.

VARIATION IN NUMBER AND ARRANGEMENT OF RAYS

While all of our native species of starfish have normally five rays of nearly equal size, yet if one examines a large number of specimens he is likely to notice certain variations. In the common starfish, for example, careful search may reveal an individual with but a single well formed ray and other rudimentary ones, while another specimen may have as many as eight rays, all perfectly formed.

Explanation of Plate VII. Variation, *Asterias forbesi.*

Fig. 1. Four-armed individual, showing no indication of injury.
Fig. 2. Eight-armed individual, provided with two madreporic plates.
Fig. 3. An individual in which one arm is forked or longitudinally divided nearly to the disk.
Fig. 4. Six-armed specimen.
Fig. 5. One with seven well formed arms.
Fig. 6. An individual in which an injury has removed all the arms of the disk except one, the four small arms being in process of regeneration.

(All figures two-thirds natural size.)

Professor Verrill, of Yale University, has made a large collection of these abnormal specimens, and has recently published* a paper on the subject. On Plate VII are shown a number of such specimens. Some of these are the result of injury, others may have arisen from congenital variation, while others evidently represent cases in which there has been a partial division of the body in an early stage of development. Fig. 2, for example, shows an individual which has eight perfectly formed rays, but this specimen is also provided with two madreporic plates. These facts would seem to indicate that in an early stage of development, perhaps even at the time of the first cleavage of the egg, the body became partially divided into two incomplete individuals. As development proceeded, these twins, as they may be called, remained attached together, each provided with four rays and the usual madreporic plate. A further and complete separation might result either in two normal individuals, identical twins, or in such specimens as the one shown in Fig. 1. On the other hand it seems possible that the four-rayed starfish of Fig. 1 may have resulted from an early injury and loss, without regeneration, of the missing arm; or as a so-called congenital variation, that is, one which has arisen very early in development, but of which the cause is not apparent. In Fig. 3 is shown an individual in which the arm opposite the madreporic plate is forked or split longitudinally from the tip nearly to the disk. This may be looked upon as a partial duplication or twinning of this ray, or possibly as the result of early injury.

Fig. 4 shows a variation in which there are six perfect rays, while in Fig. 5 there are seven. The six-rayed condition is much less rare than the other cases enumerated. From a great number of specimens which Professor Verrill has examined from near New Haven, Connecticut, 1 in about 2,000 has six rays, 1 in 3,000 has four rays, and 1 in 10,000 has seven or eight rays.† Some of these may be due to repairs of injuries received in early life, with development of extra rays.

The experimental studies of Dr. Helen D. King on the regenerative processes in *Asterias vulgaris*‡ indicate the manner

---

* *American Naturalist,* 1909.

† *Loc. cit.,* p. 546.

‡ *Archiv für Entwickelungsmechanik,* vol. ix, 1900.

in which such variations may be artificially produced. For if the disk of a starfish of this species be cut between the bases of two of the rays by an incision reaching from the border of the disk to the mouth, one or two new rays may sometimes be produced in the healing of the wound. Such supernumerary rays are at first smaller than the others, but it seems reasonable to assume that by the compensatory regulations of the body they might eventually become similar in all respects to the five original rays. Some of the six- and seven-rayed individuals of the common starfish may have resulted from similar wounds inflicted by fishes or by accident. In the removal of two or more rays, it sometimes happens that the number regenerated is one less than the normal. In this way a specimen like that shown in Fig. 1 may result.

In Fig. 6 is shown a specimen in which it is apparent that some injury has deprived the animal of all its rays except one; the four lost members are being regenerated in their normal positions about the disk. It is not at all unusual to find individuals which are thus replacing one or more rays which have been lost by an injury. Such regeneration can be studied experimentally in the laboratory, and is discussed in the following chapter.

Nevertheless it should be emphasized that, while certain species from other parts of the world naturally reproduce by spontaneous division, it is not believed that our native species can do so. As stated in the following chapter, a single ray without any portion of the disk cannot reproduce the missing parts, and there is good reason for believing that a large animal with even one or two rays removed is often destroyed by its enemies before regeneration has taken place. In smaller individuals the injury caused by amputation seems less serious. But it is only when protected by favorable conditions that the injured starfish survives in the struggle for existence.

REGENERATION

It is said that many years ago, when the starfish first began to be recognized as an important enemy of the oyster, the oyster fishermen sought to wreak vengeance on these devourers by taking such as came into the boat with the oysters, tearing them into two parts and throwing both parts overboard. In how large a proportion of cases such a mutilation resulted in the death of the starfish it is impossible to decide, but an examination of any considerable number of specimens, especially of the common starfish, *Asterias forbesi,* will show how futile are any efforts to destroy the animal merely by tearing it in two, for many individuals will be found in the process of regenerating one or more missing arms.

Such specimens indicate the readiness with which, under favorable circumstances, the starfish may repair its injured tissues, and replace one, two, three, or even four lost or mutilated arms. In fact, when carefully protected and fed in an aquarium, the disk alone, deprived of all five arms, may slowly regenerate all the missing parts.

If an arm be pulled forcibly from the disk, it often happens that a portion of the disk is removed with the base of the arm. If the arm be badly mutilated and not removed, it is usually thrown off by the spasmodic contractions of the muscles at a certain point immediately outside the disk. Very strong chemical or electrical stimuli may cause the same result.

This power of autotomy is doubtless of much importance to the starfish, which can thereby stand a chance of escaping with some of its arms when an enemy has seized one or more. Certain species living in other parts of the world, habitually reproduce themselves by this method of fission, although none of our native species multiply in this manner. In our species the disk alone can restore the arms, but a single arm cannot restore the disk.

Nevertheless it should be emphasized that whenever a starfish is mutilated from any cause it becomes at once more susceptible to the attacks of its enemies, the injured surface furnishing a vulnerable point, from which the protruding soft tissues may be seized by fishes, crustaceans, and other creatures. The whole body may be destroyed before the wounded surface has healed.

An arm removed from the body of a starfish may live for several weeks if protected from the larger carnivorous animals, but will eventually die without regenerating the rest of the body.

But the disk from which one or more arms are missing will, if not destroyed by fishes or other enemies, immediately begin the healing of the wounds and the regeneration of the missing parts.

Careful observations on this point which have been made by Mead* show how rapidly these regenerative processes may occur in the common starfish.

The removal of a single arm has little if any influence on the rate of growth of a young starfish, for such mutilated specimens may grow as rapidly as normal individuals kept under similar conditions. Futhermore the rate of growth of the regenerating arm is on the average somewhat more rapid than that of the remaining arms, so that there is a tendency to restore the normal symmetry of the body. This is shown by the fact that, of five individuals subjected to the experiment, the increase in the length of the largest arm in each during about eight weeks was 8, 10, 10, 11, and 10 mm., while the regenerating arm in the same individuals grew 12, 11, 12, 11, and 9 mm., respectively, in the same period.

The removal of two arms in a young starfish likewise has little if any effect on the rate of growth when food is abundant. In such cases also an accentuated growth in the regenerating arms as compared with the others illustrates again the compensatory properties of the body.

Although, as has been stated, it seems improbable that an isolated arm can, in the common starfish, regenerate the missing parts, yet the result is quite different if a small portion of the disk is removed with the arm. In such cases, although growth ceases for a long time, the normal body may eventually be restored from the small portion of the disk remaining.

Furthermore it is often possible to cut the disk with its attached arms into two or more pieces and from each piece obtain a perfect starfish.

From such results it is conceivable that a starfish taken from the water and torn roughly into two parts may, if a portion of the disk remains attached to each part, eventually form two complete animals by regeneration. That such an event is of usual occurrence, however, is very improbable, because of the enemies which may destroy the injured parts. It is far more likely that

* *Bulletin U. S. Fish Commission*, vol. xix, 1899.

the part having the greater portion of the disk will quickly regenerate its missing organs, while the other part will perish.

The regenerative capacity of the northern starfish, *Asterias vulgaris,* has been the subject of an extensive series of experiments by Dr. Helen D. King.* In this species, as in the common starfish, a single arm without any portion of the disk may live for several weeks, but is unable to reproduce the missing parts of the body.

Of nearly two thousand individuals of this species examined, about eleven per-cent showed one or more arms in process of regeneration. In all cases except one the regeneration involved the production of one or more new arms from the disk, and not a repair of the injured tissues of an arm. It is thus assumed that a badly injured arm is in nature thrown off and replaced by a new one.

When the tip of an arm is removed, however, regeneration of a new tip occurs, and the new eye-spot may be formed within a week after the operation. The regeneration of a complete arm from the disk requires several months.

When the arms are split longitudinally they are always thrown off, except when the tip alone is split. In the latter case two new tips may develop, giving a forked extremity to the arm. Oblique cuts and the removal of small pieces result in the normal regeneration of the injured tissues. The wounds made by cutting the disk may lead to the formation of supernumerary arms, as described in the preceding chapter.

REPRODUCTION

The starfishes are of separate sexes, although there are no external characters by which the sex is indicated.

One of the chief reasons why the starfish occurs in such enormous numbers, is found in the immense number of eggs produced by a single female. The animals are capable of breeding, as has been recently found by Mead, if the food supply is abundant, when but one year old. As the animal increases in size during succeeding years, an increasing number of eggs or sperm cells are produced annually.

* *Archiv für Entwickelungsmechanik,* vol. ix, pp. 724-737, 1900.

The sexual glands vary greatly in size according to the degree of maturity of the sexual products. They are voluminous, as shown on Plate V, and occupy, when mature, much of the space within the body walls. The number of eggs produced by a single individual in one year varies with the size of the animal, and is also largely dependent upon the previous food supply.

The genital organs develop in intimate relation with the tissues adjacent to the madreporic canal. When fully formed they consist of five pairs of voluminous feather-like bodies situated in the interradial spaces of the body, and extending freely into the arms, as shown in Plate V. Each gland opens to the exterior by a minute pore on the aboral surface near the base of the corresponding side of an arm.

When the sexual products are fully ripe, the eggs of the female pass out of the genital pores and float about in the water or sink slowly to the bottom.

The males at the same season discharge from similar openings milky clouds containing millions of minute sperm cells. Each of these sperm cells is provided with a vibratile tail, by means of which it swims about in the water. Whenever one of these sperm cells comes in contact with an egg recently discharged from the female, it bores its way into the substance of the egg, and fertilization results. Only a single sperm cell normally enters an egg; if others approach, their entrance into the egg is usually inhibited. If more than one actually succeeds in penetrating the egg substance, abnormal development occurs. Such eggs as are not fertilized within a short time after being deposited perish.

After fertilization a series of internal changes occurs resulting in the fusion of the nuclei of egg and sperm. The processes of cleavage and embryonic development then ensue, as described in the following chapter.

The season at which the sexual products are matured in Long Island Sound varies considerably with different individuals. When examined in the fall the sexual glands of all the small individuals and of many of the larger specimens are small and inconspicuous. During the winter they increase in volume and by early spring have attained nearly full size. Their feather-like processes then extend far out into the cavities of the arms, and they have taken on a rosy color characteristic of approaching maturity. Yet, in spite of the apparent ripeness of the sexual glands, the sperm cells of many individuals lack the activity of full development, and the eggs of the females when examined under the microscope reveal their immaturity.

In May the sexual glands have become so greatly distended in well-fed individuals that the arms appear full and rounded. The feather-like processes of the glands have now become so delicate that a slight injury is sufficient to allow the sexual products, eggs or sperm cells, according to the sex of the individual, to flow out freely into the water. From this time on through the following month the eggs of a majority of specimens can be easily fertilized artificially, and make extremely interesting objects for the study of cleavage and early development. Owing to the minuteness of the sexual products, however, a compound microscope is necessary for the study.

To obtain the eggs for study plenty of clean sea water must be at hand together with suitable receptacles of glass or porcelain. Soup dishes or common tumblers are very convenient, but must be kept covered. The starfishes must, of course, be kept alive in sea water, and in order to insure ripe products of both sexes a considerable number of specimens should be available.

After placing clean sea water in one of the vessels a starfish may be taken and, by means of strong scissors, opened by removing the aboral surface of one or more of the arms. The feather-like glands may be taken with a pair of forceps and shaken gently in the vessel of water. A milky cloud issuing from the cut end of the gland indicates that the specimen is a male, the cloud itself being made up of thousands or millions of sperm cells. If, on the other hand, instead of a milky cloud, there issues a stream of tiny particles barely visible to the naked eye, the animal opened is a female, and the little particles are the eggs, or ova. Examination of a drop of the contents of the vessel under a microscope will prove definitively which sex has been obtained. One or more other starfishes must be opened in a similar manner, the generative products of each being received in a separate vessel, until the genital products of both sexes have been obtained.

After the eggs in one of the vessels have settled to the

bottom, the greater part of the water should be poured off and be replaced by clean water. No more eggs should be left in the vessel than will cover the bottom in a single layer. A drop or two of the water in a vessel containing the sperm should now be added to the eggs, and will probably contain a sufficient number of sperm cells to fertilize all the eggs present.

Within an hour and a half after fertilization, the eggs, which should be examined from time to time under the microscope, will have divided into two cells. The four-, eight-, twelve-, and sixteen-celled stages follow rapidly, and resemble the cleavage stages of the sea-urchin shown in Fig. 15.

In about twelve hours more, if the temperature is fairly warm, free-swimming ciliated embryos will be found. If a few embryos are kept in plenty of clean, cool sea water, they will live for several days and develop into the so-called brachiolaria stage, as shown in Fig. 4.

Under the normal conditions of the sea, the free-swimming stages last about three or four weeks, after which such larvæ as have survived settle to the bottom and attach themselves to rocks, seaweeds, eelgrass, or other objects, and by a complicated metamorphosis become transformed into tiny starfishes.

As has been stated above, the majority of individuals of the common starfish mature their genital products in May and June, and discharge them into the water during the latter month. There is, however, great individual variation in this respect, for an examination of a large number of specimens at any season of the year, even in the fall, will show some in which the genital glands are large and contain ripe or nearly ripe sexual products. Furthermore, during the months immediately following the normal spawning season there are some individuals which are filled with mature genital glands. Such individuals apparently represent instances of belated maturity of the sex products, or those which for some reason have failed to discharge their ripened germ cells at the normal time.

Such individuals, however, probably have little influence in propagating the species; for, even if their sexual products are discharged into the water, there is very little chance that fertilization can occur at other times than in the normal spawning season, when the water is swarming with the germ cells of both sexes.

Because of the vastness of the volume of the surrounding sea water it is only when both sexes discharge their products within reasonable proximity to each other and at very nearly the same time that the eggs have any chance of being met by the sperm cells. Nevertheless, under favorable conditions of currents in the water, the sperm cells may be carried to eggs deposited at some considerable distance, probably several hundred feet.

Except in rare instances, therefore, such individuals as fail to mature their sexual cells at the normal period and discharge them into the water in the vicinity of one or more individuals of the opposite sex give rise to no offspring during that season. Moreover, it is uncertain as to whether such sexual products as mature at unusual seasons are discharged from the body.

The northern starfish, *Asterias vulgaris,* exhibits breeding habits similar to those described for the common starfish. In this species also the sexual products mature in summer.

The breeding habits of the blood starfish (*Henricia*) differ widely from those of the two preceding species of *Asterias,* in that there is no free-swimming larval stage. The eggs of this species are large and heavy, being filled with an orange-colored yolk. But few eggs are produced by each female. They give rise to embryos which pass though an abbreviated larval development, and quickly assume the form of the adult. Fisher has recently described* the brooding of the eggs in the case of one of the Pacific coast varieties of this species. The brood pouch is formed by arching the disk and bringing the bases of the rays close together. Into this cavity the eggs are tightly packed, and the developing young are thus protected by the mother until after the adult form has been reached.

The slender-armed starfish (*Asterias tenera*) is stated to brood its eggs and developing young in the same manner.

DEVELOPMENT

Common Starfish. — The eggs of the common starfish, as stated in the preceding chapter, are set free in the surrounding sea water mainly during the month of June. The developmental processes are very complicated, and many of the details are not

* *U. S. National Museum, Bull. 76,* p. 277, 1911.

yet well understood. A general account of the larval stages is given in the earlier papers by Agassiz,* and by Mead.†

The eggs fall to the bottom, and after fertilization the process of cleavage takes place, as described for the egg of the sea-urchin (page 100). In about twelve hours the substance of the egg has been subdivided into hundreds of cells which arrange themselves to form the body of an embryo. The surface of the body becomes covered with hair-like processes, termed cilia, which are kept in constant vibration and enable the embryo to leave the bottom and swim free in the water.

The body of the embryo becomes further differentiated, and a mouth, digestive system, and other organs are formed. The embryo can now take into its body still smaller organisms than itself for food, and under favorable conditions increases rapidly in size. Meanwhile the rounded shape of the body has changed into a structure with a number of projecting lobes covered with very long cilia, and the embryo is now termed a *bipinnaria.* In this stage the embryos are very sensitive to light and atmospheric conditions, and collect in vast swarms which at times float near the surface and at other times sink deeper into the water. They may be carried back and forth by the tides, winds, and currents, and become widely separated from their birthplace. They have a considerable power of locomotion, but do not swim for a long time in any given direction.

By further modifications the embryo develops into the strange-looking, transparent creature shown in Figs. 4 and 5. In this stage, known as the *brachiolaria,* the body is provided with several pairs of long arms covered with vibratile cilia. As the brachiolaria bears no resemblance whatever to the adult starfish, naturalists for a long time supposed it to be an entirely different kind of animal. And yet from such a creature the young starfish develops by a complicated metamorphosis.

After swimming about in the water for some three or four weeks, depending largely upon temperature and other external conditions, the larva becomes full-grown, and the brachiolaria

* On the Embryology of Echinoderms. *Proceedings Am. Acad.*, 1863; and *Mem. Am. Acad.*, 1864. North American Starfishes. *Mem. Museum of Comparative Zoölogy*, 1877.

† The Natural History of the Starfish. *Bull. U. S. Fish Commission* for 1899.

is now some fifty times as large as the original embryo. It is now ready for the metamorphosis.

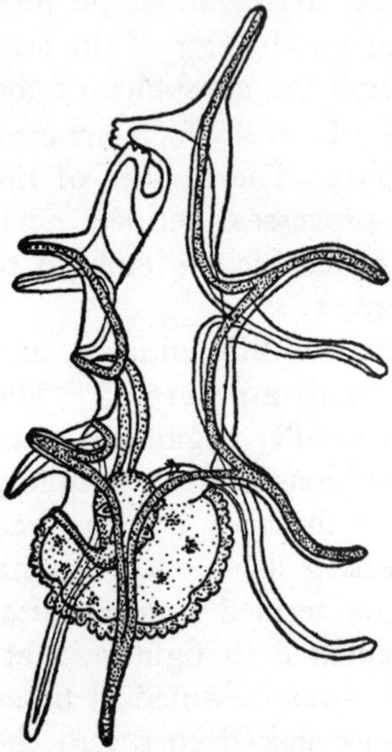

FIG. 4. Free-swimming larva (Brachiolaria) of starfish. (After Weysse.)

The wavy disk-like area at the lower end of figure represents the portion of the body which will remain after the transformation to the adult stage. Compare Figs. 5 and 6.

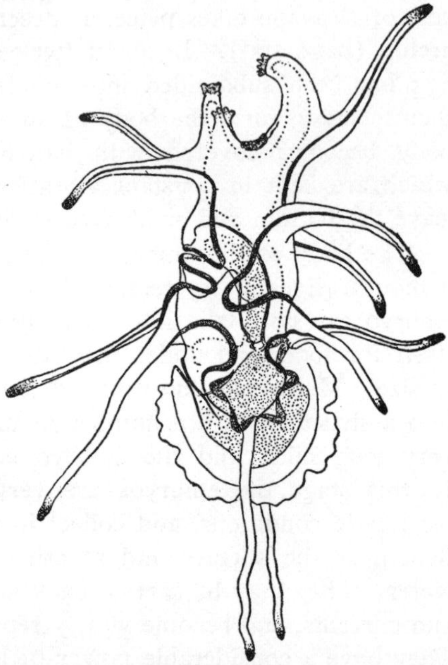

FIG. 5. Free-swimming larva (Brachiolaria) of starfish. (After Mead.)

The stippled areas in the central and lower parts of the figure indicate the position of the digestive tract, the mouth being shown at the left-hand edge of the figure, with the intestinal opening below.

The external changes by which the metamorphosis is accomplished, and the habits of the young starfish, are described by Mead* as follows:

"Already the rudiment of the resulting starfish can be seen within the brachiolarian. The five crenate lobes on the margin of the disk are the beginning of the five arms. The disk itself at this time is already somewhat opaque.

"When the larva is about to 'set,' it attaches itself to some object, like a spear of eelgrass, by the suckers, shown at the top of the figure (Figs. 4 and 5), and then a rapid transformation

* *Bull. U. S. Fish Commission* for 1899.

occurs. The whole superstructure above the disk collapses and becomes absorbed like the tail of a tadpole. In a few hours the brachiolarian has disappeared, and a starfish proper (Fig. 6) has taken its place.

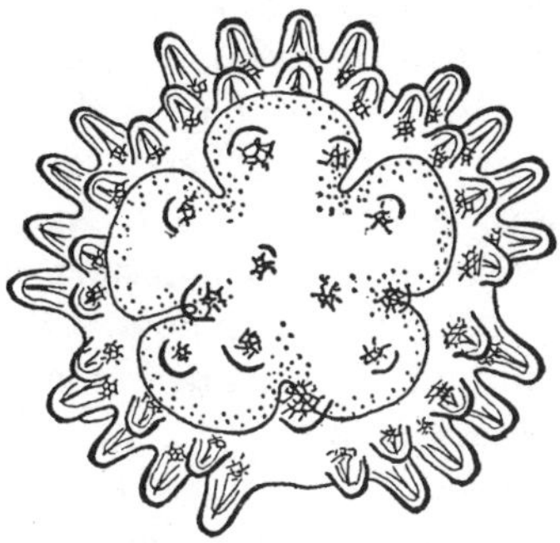

FIG. 6. Aboral surface of starfish shortly after the metamorphosis. (After Brooks.)

"Up to the very time when the larvæ are ready to set they swim freely in the water; and larvæ, caught in the tow-net, often set in the dish of water before I had returned to the house-boat, *i. e.*, within an hour of the time they were caught. In this condition they attach themselves by their suckers to any object they happen to strike, and cling to it with great tenacity until the metamorphosis is completed. As the larvæ are borne along by the currents, the eelgrass, rockweed, and especially the fluffy, branching seaweed, naturally catch immense numbers of them. I think it would not be an exaggeration to say that on a single handful of seaweed which I picked up about the first of July there were more than a thousand young stars. For the next three weeks they remain for the most part crawling about over this vegetation, gradually working down among the roots of the rockweed and upon the large stones at the bottom. They grow rapidly during this time, but decrease in numbers, for they are bright and conspicuous objects for the small fishes; yet they are exceedingly numerous for a long time. In order to obtain a definite expression of their abundance, I scooped up a large handful of the fluffy seaweed, which, together with the water, filled about two-thirds of a paper pail, and from this 603 young stars were taken. A cart-load of seaweed taken out at this time would have destroyed millions of starfish.

"By the first of August the fluffy, branching seaweed, which bore so many young stars, was nearly all dead, and though the stars were still present in great numbers upon the eelgrass, rockweed, and stones covered with sea-moss, they were also frequently seen crawling along the muddy bottom. By August 15 the eelgrass was overgrown and lodged by a luxuriant growth of *Botryllus,* a compound ascidian, which appears as dark gelatinous patches. The small stars were still numerous upon it, but were rather thin and poor. The larger and better nourished stars had left the eelgrass and were searching for food upon the stones and along the bottom.

"The small starfishes, such as live upon the eelgrass, are remarkably hardy in some respects. They will live for weeks, and even months, in a small dish, without change of water and with a minimum amount of food. During the first week in July I carried a number of free-swimming brachiolaria to Providence for further examination. They were in a glass 1-quart jar, and, after one or two were taken out, the jar was closed and was left unopened during the rest of the summer. In a few days the larvæ had all set, and when I examined the dish again, on September 5, it contained still a few live stars, which were, however, very small. Upon watching them it was seen that the more enterprising individuals were eating their companions, and finally only one remained. This one lived in the jar for weeks, but, unfortunately, I am not able to record the exact date of his death.

"On the other hand, the same young starfishes, which can live so long without food or change of water, perish quickly if left out of the water, especially if the sun is shining. They cannot live, therefore, above the low-water mark, unless sheltered by a dense growth of vegetation. Large stars can endure very much longer exposure, since their bulk prevents their drying so quickly. On July 16 I made a special search for young stars on the seaweed, above the low-water mark. I found none, yet just below low-water mark they were excessively abundant. At the same time it was noticed that above the line where the starfish were abundant there was a thick set of 1-year-old oysters, while below it the oysters were absent. The oysters set somewhat later in the season than the starfish, and the latter, therefore, are ready to prey upon the young oysters as soon as they appear. When, in addition to these facts, we take into account the extraordinary voracity of the young starfish, their immense numbers, and their special fondness for oysters, we are led to conclude that one reason why a considerable set of oysters is so rarely obtained below low water is that they fall prey to the starfish. The oysters which set above low water are comparatively safe, for when the tide leaves them uncovered they can endure for hours the direct heat of the sun, which would kill the young starfish in a few minutes.

"While the starfish are living upon the eelgrass and seaweed they are supplied with an abundance of food in the form of the young of marine worms, snails, and other animals, which, like the stars themselves, swim freely in the sea for a time, and then settle down upon any object with which they happen to come in contact. Throughout July the water at Kickemuit was teeming with minute free-swimming creatures, and in the aquarium the growth of the youngest stars could be greatly accelerated by feeding them the contents of the tow-net. During the last four days of June innumerable larvæ of a marine worm (*Syllis*) were swarming at the surface, and on July 11 millions of the young of one of the sea snails (*Littorina?*) were caught in the tow-nets.

"The clam, also, is one of those unfortunate animals whose larvæ set at about the same time as the starfish, and in the same places. The starfish before they are three days old show a predilection for young clams, which apparently does not diminish so long as any clams remain."

In Fig. 7 one of these young starfish is shown in the act of devouring a small clam. A single young starfish has been known to devour as many as 56 small clams within a period of six days.

"From the foregoing it appears that the starfish set for the most part during the last few days in June and the first week in July, some as late as July 16. They remain upon the seaweed in immense numbers until about the 1st of August, when many of them are found upon the bottom. By August 15 the greater portion of the stars have left the seaweed and gone to the bottom.

"The stars could be destroyed by hundreds of thousands in July by collecting and drying a few cartloads of seaweed taken below low-water mark. After the first week or two of July the collection of seaweed would do no injury to the clams."

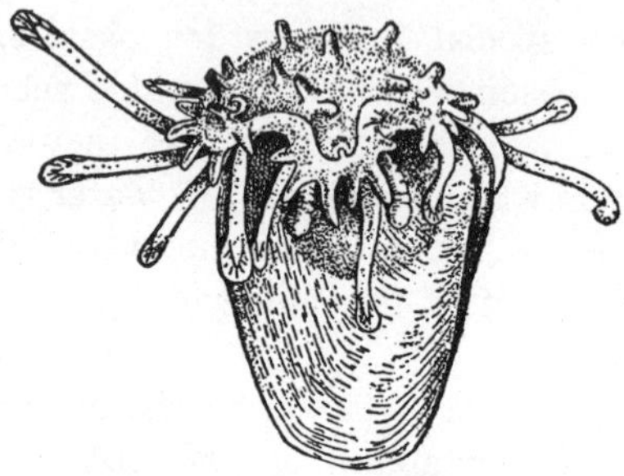

FIG. 7. Very young starfish in the act of devouring a mollusk. (After Mead.)

The rate of growth of the young starfish is discussed on page 40.

Northern Starfish. — The larval stages of the northern starfish have been studied by Agassiz*, Field†, and others, while Goto‡ has made detailed investigations on the metamorphosis. According to these observations the development of this species is similar to that described for the common starfish.

Slender-armed Starfish. — This species is reported as producing a small number of comparatively large eggs, which develop without free-swimming stages. It is also said to protect its eggs and embryos beneath the peristome until the young starfishes have reached a diameter of several millimeters.

Blood Starfish. — In this species only a few large eggs are produced. These develop into the adult form by a complicated metamorphosis, although, as stated on page 52, the larvæ do not swim free in the water.

KEY TO SPECIES

The starfishes of the world may be divided into five orders, only two of which are found in Connecticut waters or in their

* *Mem. Museum Comparative Zoölogy*, 1877.

† *Quarterly Journ. of Microscopical Science*, 1892.

‡ *Journ. College of Science, Imp. Univ. Tokio*, 1898.

vicinity. The order **Spinulosa,** represented by the beautiful little blood starfish, *Henricia sanguinolenta,* includes the more primitive forms, in which the disk and rays are everywhere thickly covered with minute spines, but there are no pedicellariæ. In this order there are but two rows of tube-feet. The order **Forcipulata** includes our three species of *Asterias.* In this order the disk and rays have scattered coarse spines, and very numerous pedicellariæ. There are usually four rows of tube-feet.

Our native species may be distinguished by the following characters:

1. Disk and rays comparatively smooth, covered with minute, closely set spines; pedicellariæ absent; tube-feet in two rows ....**Henricia sanguinolenta**

   Disk and rays rough, with coarse spines; forceps-like pedicellariæ abundant; tube-feet in four rows............2

2. Rays nearly cylindrical, tapering and slender; spines very numerous, slender, sharply pointed, usually surrounded with a broad wreath of small pedicellariæ.............. **Asterias tenera**

   Rays broad, somewhat flattened; spines thick and blunt....3

3. Rays blunt at the ends; skeleton quite firm; spines scattered; pedicellariæ near ambulacral grooves short and broad; madreporic plate usually orange........**Asterias forbesi**

   Rays pointed at the ends; skeleton rather soft; spines scattered, but often forming a rather distinct median longitudinal row on the aboral side of each ray; pedicellariæ near ambulacral grooves slender; madreporic plate pale yellow ....**Asterias vulgaris**

Asterias forbesi (Desor)

Common Starfish

Plate I, fig. 4; Plates II, III, IV, VI, VII, VIII.

This well-known pest of the oyster is found along the Atlantic coast from the Gulf of Mexico to Maine, and is the only species of starfish which is common in the greater portion of Long Island Sound.

The species is so generally distributed in the Sound that the mention of specific localities would be superfluous. As a rule it is most abundant on shelly bottoms in shallow water. In the fall the tide pools in rocky localities along the shore are often crowded with young starfishes measuring from a half-inch to three inches between tips of opposite arms. These are feeding on the barnacles and young oysters which are attached to the rocks. At the approach of winter most of the starfish leave the shore and migrate into deeper water.

In spite of the thousands of bushels of these starfishes which are annually destroyed by the oyster growers there seems to be no diminution in their numbers except in restricted localities.

The size of average mature individuals of this species is from four to six inches between the tips of the opposite arms, but much larger individuals are sometimes found. The variation in the number of rays is discussed on page 43. The shape of the rays depends somewhat on the state of contraction of the ambulacral groove, and varies with the state of development of the sexual glands and the general nutrition of the body. In general (as shown on Plates II, III, IV, and VIII) they are thick and stout, with blunt and rounded tips.

Individuals of this species can usually be distinguished from those of *Asterias vulgaris* by the difference in the proportions of the rays, as shown on Plate VIII. The latter species usually has more slender and pointed rays, while those of the common starfish are comparatively thick and blunt.

The coloration of this species is subject to great variation, the most common shades being greenish or brownish green, with light yellow spines and orange-red madreporic plate. The colors of other specimens range through brown, bronze, orange, and purple. There seems to be little if any correlation between the color of the animal and the age, sex, or environment, although in

---

EXPLANATION OF PLATE VIII. Upper figure, *Asterias forbesi.* Three-fourths natural size. Lower figure, *Asterias vulgaris.* One-half natural size.

Photographs of dried skeletons from oral surface, to show the different proportions of the arms in the two species. The skin, tube-feet and other soft parts have been removed. Around the mouth may be seen in each figure five groups of large spines, the oral spines, or mouth papillæ. In the upper specimen the oral spines lie in the positions they occupy when the mouth is closed, while in the other specimen the mouth is fully opened. The ambulacral grooves are widely opened, and show the four rows of openings for the tube-feet in each groove. (From photographs loaned by Prof. A. E. Verrill.)

PLATE VIII. *Asterias forbesi* and *Asterias vulgaris*.

PLATE IX. Northern Starfish, *Asterias vulgaris.*

restricted localities a definite color variety may predominate. After death the body assumes a dull reddish or orange color, and the madreporic plate is yellow or brown.

The plates of the aboral surface bear numerous blunt spines surrounded by circles of pedicellariæ (Plate III). The respiratory papulæ, or branchiæ, are grouped in clusters of four to six or more in large areas between the spines. Each of the plates adjacent to the ambulacral grooves bears usually two slender, flattened, sharply truncated spines. The number of such plates increases with the growth of the animal, being from 80 to 120 in well grown individuals.

A full and interesting account of the habits and life history of this species is given by Mead, in the Bulletin of the U. S. Fish Commission for 1899.

Asterias vulgaris Verrill

Northern Starfish

Plates V, VIII, IX.

Although this species is found in the colder waters of the eastern part of the Sound, its more usual habitat is farther to the northward. It is occasionally found as far west as Faulkner's Island and the Thimble Islands, but becomes more common from New London eastward, particularly in the deeper and colder waters.

This is the common starfish north of Cape Cod, and ranges northward to the coast of Labrador. In the deeper, colder waters off the Atlantic coast the species extends south as far as Cape Hatteras. In certain localities, as at Woods Hole, Mass., Narragansett Bay, and the eastern part of Long Island Sound, this species is sometimes associated with the common starfish. The same or a very similar species occurs on the coast of Europe, where it is known as *Asterias rubens*.

EXPLANATION OF PLATE IX. *Asterias vulgaris*. About one-half natural size.

Photographs of living individuals, from aboral and oral surfaces. In the upper figure the disk is somewhat unusually swollen. The lower figure shows the four rows of tube-feet and the mouth. These figures well illustrate the characteristic arrangement of the spines on both surfaces of the body.

(Photographs loaned by the U. S. Fish Commission, with permission of Dr. H. L. Clark.)

In general it is possible to distinguish our two species by the shape of the body, the arms being thicker at the base and more pointed at the tips in the northern starfish (Plates VIII, IX), while those of the common starfish are rounded and blunt at the tip. The spines along the middle of the aboral surface of each arm often form a conspicuous median row (Plate IX) extending the length of the arm, while those of the common starfish are scattered more irregularly.

Another specific distinction may be found in the shape of the major pedicellariæ, particularly those on the adambulacral spines, as shown in Fig. 8. Instead of the broad, rounded pedicellariæ found in *A. forbesi,* many of those of the present species have long and pointed blades.

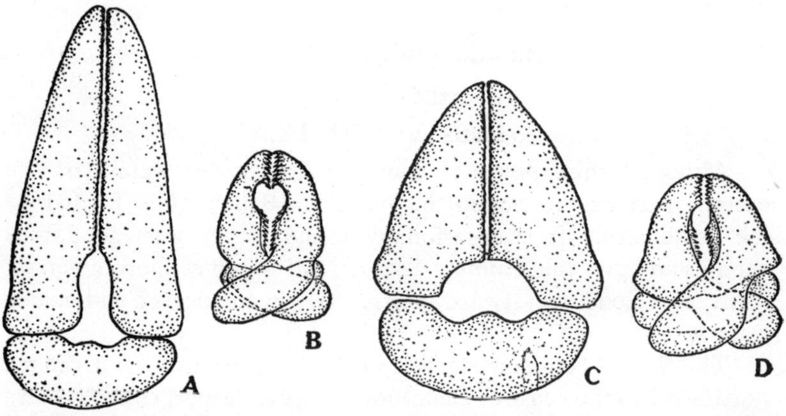

FIG. 8. Calcareous plates of pedicellariæ of *Asterias.* A, major pedicellaria, *Asterias vulgaris;* B, minor pedicellaria, *A. tenera;* C, major pedicellaria, *A. forbesi;* D, minor pedicellaria, *A. forbesi.* All from adambulacral spines. Enlarged 90 diameters.

This species grows to a much larger size than does the common starfish, full-grown individuals often measuring eight inches or more between the tips of opposite arms. Specimens more than 16 inches in diameter have been recorded from Nova Scotia.

The coloration shows great variation, yellow, yellowish brown, orange, and purplish specimens being most abundant, although shades of pink and red are sometimes present. The madreporic plate is usually pale yellow. The spines are generally of lighter color than the surface of the body.

PLATE X. Slender-armed Starfish, *Asterias tenera.*

Asterias tenera Stimpson

Slender-armed Starfish

Plate X.

In the paper by Stimpson* in which the original description of *Asterias tenera* occurs, there is described another, very similar species, under the name of *A. compta,* based on a single specimen three inches in diameter.

Asterias compta was thought to differ from *A. tenera* in being of somewhat less slender proportions, in having the rays more depressed, and particularly in having the spines, especially towards the tip of the rays, surrounded by broad wreaths of pedicellariæ. The type locality for *A. tenera* was Massachusetts Bay, and for *A. compta* off the coast of New Jersey.

From an examination of the records of the collections of this species by the United States Fish Commission, Verrill† gives the distribution of *A. tenera* as Cape Cod to Newfoundland in 10 to 129 fathoms, and states that the species is abundant in Massachusetts Bay and the Bay of Fundy in 10 to 40 fathoms.

Verrill states that *A. compta* is large and abundant in the cold area south of Rhode Island and Marthas Vineyard in 20 to 50 fathoms. He emphasizes the close relationship of the two so-called species, and suggests that *A. tenera* may be only a poorly nourished, slender variety of *A. compta.* Clark‡ has examined both varieties and considers them specifically identical.

An examination of a large number of specimens of both forms belonging to the Yale University Museum has failed to show any definite anatomical characters by which they can be invariably distinguished. Many of the specimens can be placed without hesitation in the one group or the other, but others, and

* *Proc. Boston Soc. Nat. Hist.*, pp. 269, 270, 1862.

† Distribution of the Echinoderms of Northeastern America. *Am. Journ. Science,* vol. xlix, 1895.

‡ *Bull. U. S. Fish Commission,* 1902.

EXPLANATION OF PLATE X. *Asterias tenera.*

Fig. 1. Oral surface of dried specimen showing the slender spines surrounded by dense wreaths of pedicellariæ; about twice natural size.

Fig. 2. Aboral surface; natural size.

Fig. 3. Oral surface of a small specimen preserved in alcohol, showing the broad ambulacral grooves with tube-feet; natural size.

Fig. 4. Oral surface of dried specimen; natural size.

particularly the smaller specimens, possess such intermediate characters that it has been found impossible to place them in either group. Therefore but a single species can be recognized at present, and, as the description of *A. tenera* precedes that of *A. compta* by a single page, the former has priority.

On the coast of Europe occurs a very similar species, *A. Mülleri,* and it is not certain that this is specifically distinct from our *A. tenera.* Further examination of a large series of both forms is needed to determine the matter with certainty.

Asterias tenera, including both varieties, as shown by the records of the U. S. Fish Commission, ranges from off the mouth of Chesapeake Bay northward to Nova Scotia, in from 10 to 129 fathoms. It is recorded from the eastern end of Long Island Sound, off Watch Hill, Rhode Island, but is more abundant in the colder water farther off the coast and to the northward.

The individuals of this species are small in size. Mature specimens commonly measure only from two and one-half to four inches between the tips of opposite arms. A few larger specimens have been obtained, but the vast majority of specimens in the extensive series belonging to the Yale University Museum are less than three inches in diameter.

The species may be distinguished from our other species of the genus by the nearly cylindrical, slender, pointed arms (Plate X). The disk is small, the slender arms being from five and one-half to seven times as long as the diameter of disk.

The color in life is pale, sometimes nearly white; often pale purplish or pinkish. The madreporic plate and the spines are whitish.

The rather open meshwork of plates on the aboral surface carries very prominent, slender spines (Plate X, fig. 2) which do not form a clearly defined median row on the arms, as they do in *Asterias vulgaris.* In the majority of specimens all the spines except those on the disk are surrounded by a conspicuous wreath of pedicellariæ (Plate X, figs. 1 and 4). They are more closely crowded toward the tip of the ray. In some individuals pedicellariæ occur upon the spines of the disk, while in others they are almost entirely limited to the rays. Usually one, two,

PLATE XI. Blood Starfish, *Henricia sanguinolenta.*

or a few much larger pedicellariæ are found on the oral surface of the disk in each of the angles between the rays.

When the spines of the disk are free from pedicellariæ, groups of such organs appear at the beginning of the rays near the bases of the spines, and these groups develop into conspicuous wreaths on the more prominent spines of the rays. The wreaths are most conspicuous on the large spines about the ambulacral grooves and toward the tip of the rays (Plate X, fig. 1).

The individual pedicellariæ in these wreaths, while relatively large as compared with the size of the spine, are much smaller than in our two other species of the genus (Fig. 8). They are usually oval in outline, with rather blunt points.

The madreporic plate is small, and provided with a few very wide furrows.

The branchiæ are usually separate or in groups of two in each branchial area.

As is the case with several other species of the genus, this little starfish protects its eggs and young beneath the disk, and carries about with it the young clinging to the partially everted borders of the mouth.

Henricia sanguinolenta (O. F. Müller)

Blood Starfish

Plate I, figs. 1—3; Plate XI.

This beautiful little starfish is known in the writings of most American zoölogists by the name of *Cribrella sanguinolenta.* It has recently been shown, however, that the generic name *Henricia* must take precedence over the more familiar *Cribrella.*

Like the preceding species, this is more properly a northern form, and grows far more abundantly and of larger size along the coasts of Maine and Labrador than farther south. It occurs, however, in the cold, deep water off our coast as far south as Cape Hatteras, and has been taken at a depth of more than a thousand fathoms.

Explanation of Plate XI. *Henricia sanguinolenta.*

Fig. 1. Aboral surface of disk, showing madreporic plate, intestinal opening, and arrangement of spines; three times natural size.
Figs. 2 and 3. Aboral and oral surfaces; about two-thirds natural size.
Fig. 4. Oral surface of dried specimen; slightly enlarged.

To the northward it extends not only along the Atlantic coast of North America but also to Greenland, and is found also on the northern coasts of Europe and Asia. It occurs from the Azores to the British Islands, Scandinavia, and Spitzbergen, and along the shores of the Arctic Ocean to Bering Strait. On the American side of the Pacific Ocean the species extends as far south as Washington, and on the Asiatic side southward to the Kurile Islands. This would indicate a typical circumpolar species which extends southward along the continental shores to latitude 38° to 48° N.

In Long Island Sound the blood starfish appears to be limited to the eastern part, which has deeper and colder water than is found elsewhere. It extends at least as far west as New London.

It is at once distinguished from our species of *Asterias* by having only two rows of tube-feet beneath each arm, arranged in a single row along each side of the narrow ambulacral groove (Plate XI). The proportions of the body vary greatly, as shown on Plate I. The arms are either rounded or flattened at the base, but are always rounded or nearly cylindrical distally, and are comparatively slender and acutely pointed (Plate I). In some specimens (Plate XI) the arms are swollen near the base, and sharply contracted between base and disk. Sometimes the tips of the arms are slightly raised and curved aborally, and the two arms of the bivium are brought closer together than are any of the others (Plate XI).

The aboral surface lacks the prominent spines of *Asterias*, the whole surface being comparatively smooth, but actually covered everywhere with minute closely set spines (Plates I and XI).

The branchiæ are usually scattered singly in minute spaces between the thickly placed spines. The madreporic plate is small, and marked by a few broad grooves. The anal opening is more conspicuous than in *Asterias* (Plate XI, fig. 1).

Specimens from Long Island Sound rarely exceed three or four inches between the tips of opposite arms.

The colors are unusually deep and bright, rendering the animal very conspicuous when exposed to full view. The usual color is bright red or orange on the aboral surface, although cream-colored, yellow, pink, lavender, and purple specimens are

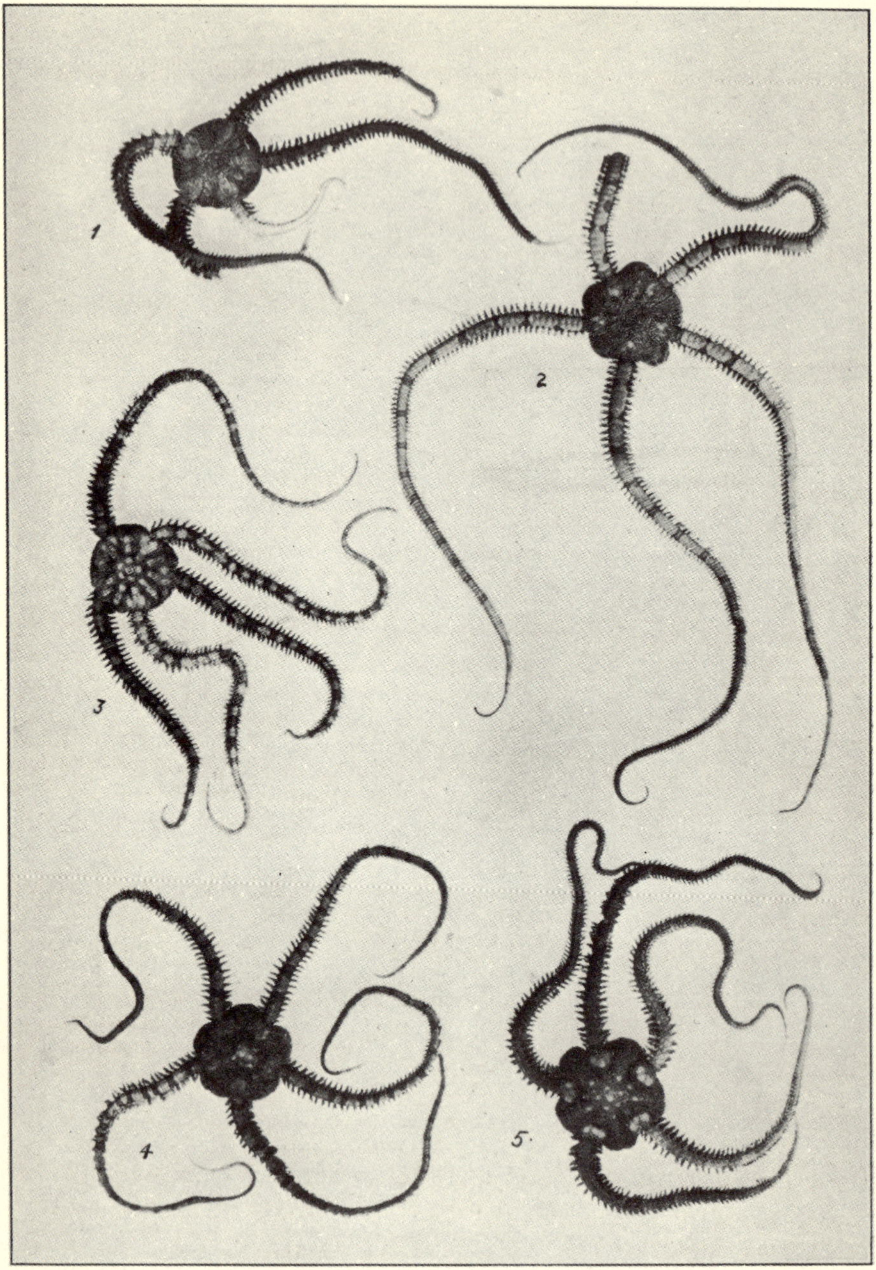

PLATE XII. Variation in form and color patterns. Daisy Brittle-star, *Ophiopholis aculeata*. (Three-fourths natural size.)

sometimes found. Occasionally the surface is mottled red and purple. It has been said that a glass dish filled with these starfishes might vie with a tulip bed in gaiety and vividness of tints. The oral surface is deep yellow.

The few large eggs laid by each female in early spring develop by a complicated metamorphosis, but the adult condition is reached without passing through a free-swimming stage. In this species the eggs and young are protected beneath the disk of the mother.

Class 2. **OPHIUROIDEA**

The representatives of this class are known popularly as serpent stars, snake stars, brittle stars or sand stars. The species which occur in the waters of Long Island Sound are small and inconspicuous as compared with the common starfish or the sea-urchin, and are seldom noticed by the ordinary visitor to the seashore. They are protectively colored, so that it is very easy to overlook them when their slender arms are stretched out among the roots of the eelgrass or coiled up in crevices of rocks. Furthermore they are of a retiring disposition and largely nocturnal in their habits, remaining secluded as much as possible during the daytime. Some forms live beneath stones or hidden among seaweeds in small crevices of the rocks. One species, *Amphioplus abditus,* lives buried in the mud, with one or more arms projecting above the surface. It is only when they are disturbed and are moving their delicate arms in an effort to escape that they are at all likely to attract the attention even of a trained observer. And, finally, although they occur in great abundance in certain localities, in most places they are so scarce or so well protected by their habits and coloration that careful search for hours at a favorable tide will fail to disclose a single specimen.

Explanation of Plate XII. Variation, *Ophiopholis aculeata.*

Photographs of five individuals to show some of the variations in form, length of arms, and color patterns of this extremely variable species. These figures represent but a few of the innumerable positions which the arms can assume in the living animal. In the upper left-hand figure all the arms are in process of regeneration. (All figures three-fourths natural size.)

STRUCTURE

The class Ophiuroidea contains hundreds of species distributed all over the world. Four of these species are known to occur in Long Island Sound, and one other has been found so near the eastern entrance to the Sound that it seems probable that further investigation will reveal its presence in Connecticut waters.

They all have a flattened, circular or pentagonal body, called the disk, with five very slender, flexible, jointed arms, or rays (Plate XII). These arms taper gradually toward the tip, and their movements in life are such as to warrant the popular appellation, "serpent stars," for, when the animal is disturbed, each one twists in all directions and folds itself about adjacent objects in a truly serpentine fashion.

The mouth is placed in the center of the disk, in a position similar to that in the starfish. Hence the side of the disk containing the mouth and the corresponding faces of the arms make up the oral surface, while the opposite side of the disk and arms is the aboral surface. The mouth is provided with five triangular jaws, each situated between a pair of arms, that is, in one of the interradial spaces (Plate XIII).

The arms are distinctly marked off from the disk, and do not pass gradually into it like those of the starfish. Ambulacral grooves are wanting, and the tube-feet are very delicate.

The disk contains all the digestive and reproductive organs, the arms containing only a row of calcareous ambulacral ossicles

---

EXPLANATION OF PLATE XIII. Skeleton of Ophiuran. *Ophiopholis aculeata.* (Four times natural size.)

Photographs of disk and bases of rays from oral and aboral surfaces to show characteristic arrangement of plates. The upper figure shows the granulations and very blunt spines which cover many of the plates of the disk, leaving only about forty of the plates exposed. In many specimens the number of exposed plates is smaller; sometimes as few as six are thus left bare. These exposed plates are usually grouped symmetrically in the five radial and five interradial areas, with a single plate in the center of the disk. The upper arm plates are seen to be separated by a row of plate-like granules.

The lower figure shows the five jaws, with four scale-like projections, oral papillæ, on each side of each jaw. In the interradial areas are five large plates, oral shields. It will be noticed that the oral shield between the arms nearest the top of the page is larger than any of the other four, and forms the madreporic plate. On each side of the ventral arm plates is a single row of flattened spines, the tentacle scales.

PLATE XIII. Skeleton of Ophiuran, *Ophiopholis aculeata*. (Four times natural size.)

with a slender branch of the body cavity, together with small nerves and branches of the water-vascular system.

The mouth leads into a narrow esophagus, and this opens into a large saccular stomach, which fills up most of the space within the disk. There is no anal opening, the indigestible remains of the food being cast out of the mouth.

Disk.— The disk is covered with thin plates on all sides. Certain of these plates present such definiteness in form and arrangement that they afford the most convenient characters for distinguishing the various species.

On the oral surface five triangular groups of plates project in toward the center of the disk and form the jaws (Plate XIII). The jaws are usually provided with numerous toothlike processes or flattened scales constituting the masticatory apparatus. These processes are known as oral papillæ, tooth papillæ, and teeth, according to the position they occupy on the jaw. The oral papillæ are toothlike projections along the edges of the jaws, while the tooth papillæ, when present, are similar structures at the point of the jaw, and the teeth are five series of plates which project from the mouth opening toward the interior of the body. The oral papillæ and teeth are well shown on Plate XIII, while Fig. 9 represents in a diagrammatic way the relations of the same plates. The number and arrangement of these processes are characteristic of each of the different species.

The jaws are provided with a musculature which permits them to be opened widely (Plate XIII and Fig. 9) or closed tightly. In the specimen shown on Plate XIV two adjacent jaws are thus closed, while the remaining angles of the mouth have remained open. In most species the jaws, teeth, and oral papillæ are not so much used in crushing the food as in separating out and swallowing the smaller or softer particles.

A large plate at the base of each jaw is termed the oral shield, while the pair of smaller plates on the face of the jaw, and beside or internal to the oral shield, are the adoral plates (Plate XIII and Fig. 9).

On the aboral surface there is great variation in the number and size of the plates, but one pair — the radial shields — lie at the base of each arm. In some species, as *Amphipholis squamata,* these radial shields are very large and conspicuous (Plate XVII), while in other species they are more or less completely covered by the more superficial scales of the disk (Plates XIII and XIV).

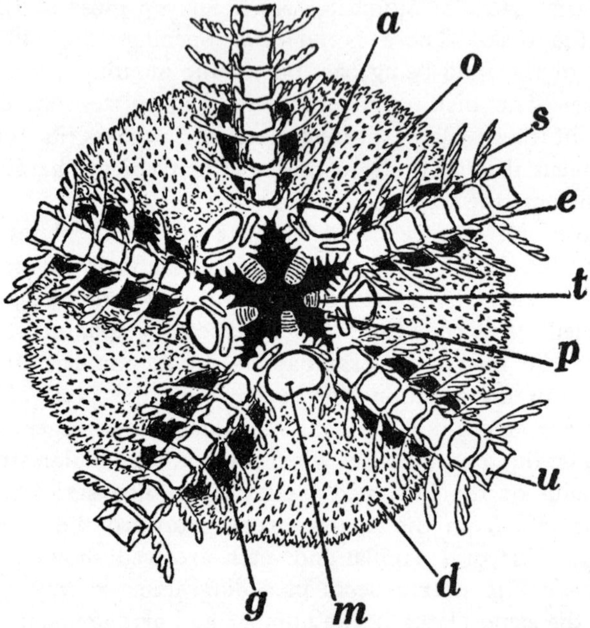

FIG. 9. Oral surface of disk and bases of arms of *Ophiopholis aculeata; a,* adoral plate; *d,* disk, with small, blunt spines covering the plates; *e,* tentacle scale; *g,* genital slit; *m,* madreporic plate (modified oral shield); *o,* oral shield; *p,* oral papillæ; *s,* arm spine; *t,* tooth; *u,* under arm plate.

Arm plates.— The arms are covered on all sides by a complicated system of thin plates and spines. The form, number, and arrangement of these parts are used in the classification of the group, so that one must examine them closely in order to determine the name of any one of our native species. A single row of plates, the upper arm plates, usually lies along the aboral surface of each arm; and a similar row, the under arm plates, often with plates of different size, occupies a similar position on the oral surface (Fig. 9). Beside these upper and under arm plates, and occupying more or less of the lateral surfaces of the arm, often occur one or more series of smaller plates, and these

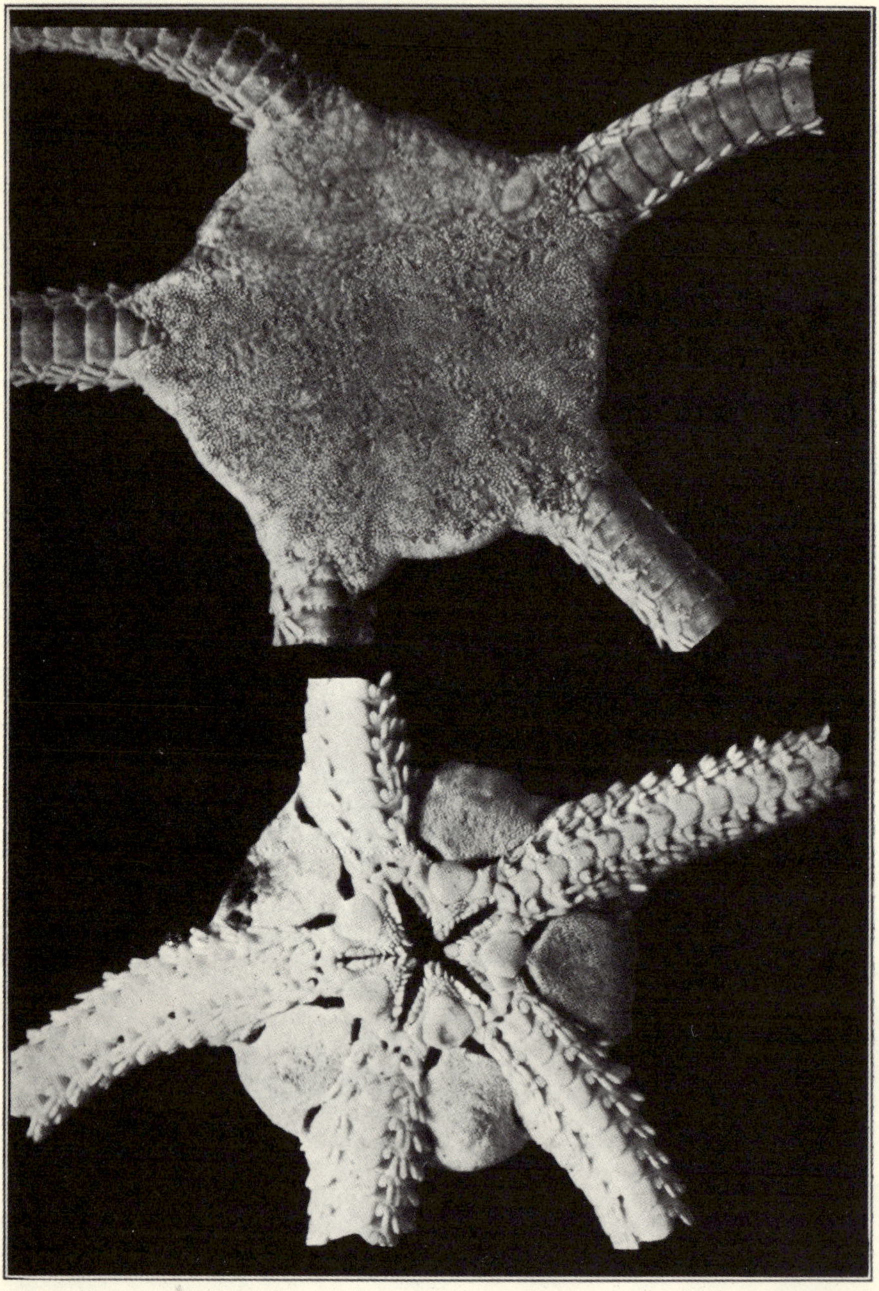

PLATE XIV. Skeleton of Ophiuran, *Ophiura brevispina.* (Five times natural size.)

commonly bear conspicuous spines, termed arm spines. The number of spines is nearly constant on the corresponding parts of the arms in all individuals of the same species, and moreover each species presents its own peculiarities in the size, position, and arrangement of these structures, as is shown on Plates XIII and XIV. Hence their importance in the diagnosis of the species.

Tube-feet.—The tube-feet are small and inconspicuous, and, as they have no sucking disks nor ampullæ, they are of comparatively little importance in the locomotion of the animal. They function largely as sense organs, and are therefore often called tentacles. They are also of service in respiration by the exchange of gases which takes place between their fluid contents and the external sea water, and in some cases they aid in passing food particles to the mouth. In certain species the greater portion of the food consists of small organisms which are secured by the tentacles as the slender arms are swept back and forth on the surface of the sand or mud. The food secured is passed along from one tentacle to the next until, on reaching the proximal tentacles, it is pushed between the jaws and into the mouth.

The tentacles protrude between the lateral plates of the arm, and are protected by delicate scales, called the tentacle scales (Plates XIII and XIV, and Fig. 9).

Internal structure of Arms.— The arm of the ophiuran is morphologically equivalent to that of the starfish, but the structures have undergone great modification. Instead of a broad

---

EXPLANATION OF PLATE XIV. Skeleton of Ophiuran, *Ophiura brevispina.* (Five times natural size.)

Photographs of dried specimens, showing aboral and oral surfaces of disk and bases of rays. In the upper figure the granules covering the surface of the disk have been removed on the side toward the top of the page to show the underlying radial shields and other plates.

The lower figure shows the plates around the mouth and on the under side of the rays. Of the five large oral shields the one on the lower right-hand side is somewhat larger than the others, and is the madreporic plate. The adoral plates lie close beside the oral shields. The figure shows the teeth at the apex of the jaws, and the seven oral papillæ on each side of each jaw. The left side of the mouth shows the position of the jaws when closed. The two pairs of large openings near the base of each arm are the genital slits. On the arms the upper and under arm plates, the tentacle scales, and the closely-appressed arm spines are shown.

central cavity occupied by the digestive and reproductive organs, the body cavity in the arm of the ophiuran is represented only by a narrow canal (Fig. 10). Each arm is made up of a great number of similar segments. The bulk of the arm consists of a series of solid, cylindrical disks, called vertebræ, which are arranged end to end in a single row, and taper gradually to the tip of the arm.

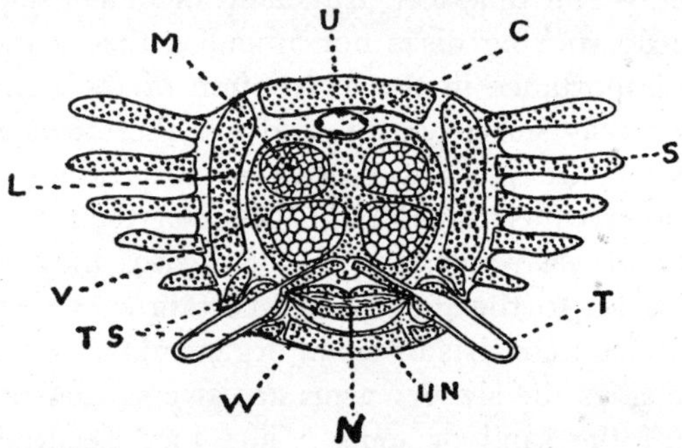

FIG. 10. Diagram of transverse section of arm of an ophiuran. *C*, branch of cœlom; *L*, lateral arm plate; *M*, muscular attachment of vertebra; *N*, radial nerve; *S*, arm spines; *T*, tentacle; *TS*, tentacle scales; *U*, upper arm plate; *UN*, under arm plate; *V*, vertebra; *W*, radial canal of water-vascular system. (Modified from MacBride.)

Each vertebra, in all our native species, articulates with its neighbor by a ball-and-socket joint, the concavity on the proximal side interlocking with a corresponding projection on the distal side of its neighbor.

The vertebræ are joined to each other by four strong muscles (Fig. 10), attached symmetrically to their faces. The contraction of the two upper muscles of a series of segments bends the arm upward; a similar contraction of the lateral or lower pair bends the arm sidewise or towards the oral surface as the case may be. By suitable combinations of such muscular contractions the arms can be moved rapidly and bent about in all possible directions, with the one exception that they are so articulated that they cannot be coiled in a spiral above the disk.

On the aboral side of each vertebra (Fig. 10) is a shallow groove which contains the slender extension of the body cavity, mentioned above, although no part of the digestive or other systems extend into it.

A similar groove on the oral side of the vertebræ contains the radial canal of the water-vascular system, which is connected with each of the tentacles, as with the tube-feet of the starfish. Beneath the radial canal is the radial nerve cord and the perihæmal canal comparable to those of the starfish.

The upper arm plates cover the extension of the body cavity, just as the under arm plates cover the organs beneath the vertebræ (Fig. 10).

Viscera.— The visceral anatomy of the ophiuran is not essentially different from that of the starfish, except that the digestive and reproductive organs do not extend at all into the rays. The simple digestive organs, mouth and saccular stomach, have been described above. The water-vascular system consists, as in the starfish, of a circumoral vessel and five radial canals, connected with the tentacles of the arms. The madreporic canal is connected with the exterior by one or more minute pores situated on one of the oral shields.

The nerve ring and its branches and the perihæmal vessels have much the same situation as in the starfish, although the nerves are more highly modified by the segregation of the nerve cells into more definite centers, or ganglia.

HABITS

The ophiurans are usually nocturnal in habit, some species hiding away in protected places during the daytime, others remaining buried in the mud. When disturbed they hastily swim or creep away. In case the irritation has been too severe some species spasmodically throw off and leave behind one or more of their arms. This causes them only a temporary inconvenience, however, for such mutilated individuals rapidly regenerate the missing parts.

Ambulacral grooves, so well developed on the oral surface of the rays in the starfish, are lacking, and the tube-feet are so little developed that the movements of the animal are controlled by the flexible, muscular arms. They provide, however, a much more rapid means of locomotion than is afforded by the tube-feet of starfishes. The ophiuran can move rapidly among seaweeds, pebbles, or other objects by the serpentine movements of these arms, and can also swim free in the water by moving a pair of arms much as we would move our own arms in swimming. When placed upon its aboral surface the animal can quickly recover its normal position.

In ordinary locomotion one or more of the arms are clasped about an object and the rest of the body drawn forward to that position, when the same or adjacent arms reach out again to an advanced position and the animal moves rapidly forward. The more posterior of the arms help push the body along at the same time that those in front are pulling.

There is no definite orientation of the body, and any of the arms may go in front, so that when the animal desires to change its direction it does not turn the body, but merely advances an arm which was previously in the rear. It may thus go in any direction without turning the body.

When spines are well developed along the sides of the arm they aid materially in furnishing a means of attachment to any desired object. The rudimentary tube-feet, or tentacles, also aid the arm in keeping its hold, although they are not provided with the sucking disks present in those of the starfish and sea-urchin.

The motion of the arms is controlled by the circumoral nerve ring. When this ring is cut, the movements of the adjacent arms are impaired; and, when it is cut between each pair of adjacent arms, coördinated movements are entirely destroyed.

The habits and response to stimuli of these animals have been the subject of much study in recent years. Cowles has recently shown* that the tube-feet are capable of selecting food materials with much precision and passing them from one tube-foot to another until the mouth is reached. Even when removed from the body the arm still reacts to food, and nutritive particles are carried from tip to base of the severed arm, while inorganic objects are seized and quickly dropped.

The ophiuran reacts positively to solid vertical walls, and the effect of such a stimulus is retained for some time, for when an animal is removed a short distance from its hiding place it moves in the direction of the same spot without error. This

* Stimuli Produced by Light and by Contact with Solid Walls as Factors in the Behavior of Ophiuroids. *Journ. of Exp. Zool.*, vol. ix, 1910.

so-called memory of past stimulation is lost under the influence of a much stronger stimulus, such as bright light, from which the animal retreats rapidly.

The behavior of the ophiuran when subjected to a great variety of artificial conditions has been studied by Glaser.* The animal's ability to solve to its advantage many of the problems set for it, such as removing rubber bands from its arms, for example, illustrates a high degree of nervous coördination.

Regeneration.— The rays of all our species are fragile, and are so likely to be broken off in handling, that in some species it is difficult to kill and preserve a specimen without mutilation. Some of the species voluntarily throw off one or more rays at a point close to the disk if the animal is even slightly irritated by rough handling or by being placed in a fluid not to its liking. Other forms when disturbed break off the arms piece by piece until nothing of the animal remains except the disk.

When one or more rays are accidentally or voluntarily broken off in a state of nature, the wound quickly heals and new rays grow out from the disk in place of those destroyed, as in the case of the starfish.

If a ray is broken at some distance from the disk the end is restored to the normal length. Individuals are often found, as illustrated on Plate XII, which show such rays in process of regeneration. If the disk is cut in two each part is capable of regenerating a complete animal.

Certain species, such as *Amphioplus abditus,* have the remarkable habit of throwing off the disk when strongly stimulated, leaving only the mouth frame attached to the arms. Under favorable conditions the missing parts can be regenerated. In *Amphipholis squamata,* in which the young develop in the genital bursæ, this process of throwing off the disk is said to be the normal method by which the young are discharged.

FOOD

Because of their secluded and nocturnal habits it is difficult to determine the exact nature of the food on which the ophiurans subsist. The animals are seldom seen to take any food whatever

* Movement and Problem Solving in Ophiura. *Journ. of Exp. Zool.,* vol. iv, 1907.

when kept in an aquarium, but an examination of the contents of their stomachs often reveals fragments of the indigestible parts of small crustaceans and other animals. Since it seems doubtful if the ophiuran would be able to capture these animals alive, it is believed that it acts mainly as a scavenger, taking into the large saccular stomach all sorts of dead animal matter and digesting out the substances capable of assimilation. Those parts which cannot be digested are thrown out of the mouth opening.

Probably the greater part of the food of many species consists of the superficial layer of organic matter which covers the rocks and floor of the sea in all situations not accessible to waves and currents. The diatoms and other minute organisms contained therein are passed toward the mouth by the tube-feet of the arms.

Some species crawl about on the sea-bottom sweeping the arms back and forth in search of food materials. Other species, such as *Amphioplus abditus,* lie for a time buried in the mud with one or more of the extremely slender arms protruding above the surface of the mud. The tube-feet (tentacles) of these arms select the food particles within reach and pass them to the mouth.

REPRODUCTION

The reproductive organs consist ordinarily of five pairs of genital glands, situated on the walls of five sacs called the genital bursæ, which open to the exterior of the body by five or ten pairs of large slits situated on the oral surface of the disk close beside the bases of the rays. These openings are called the genital slits, and are usually quite conspicuous (Fig. 9 and Plate XIV).

The sexes are usually separate, all the glands in one individual producing eggs and in another spermatozoa. In the little *Amphipholis squamata,* however, one of the glands of each genital bursa produces eggs while the other forms sperm cells.

The sexual products when ripe break out of the sexual glands into the genital bursæ and are usually discharged directly through the genital slits, although in the above-named hermaphrodite species the eggs are fertilized and develop within the genital bursæ.

Owing to the nocturnal habits of the animals the eggs of at least some of the species are discharged in the evening; that is, between sunset and midnight. They are set free in the water and are there fertilized by spermatozoa from another individual. After fertilization the egg develops into a free-swimming ciliated larva, which by a complicated metamorphosis gives rise to the young ophiuran as described in the following chapter.

The genital bursæ are thin-walled sacs; and, as the sea water contained in them has free access to the exterior through the genital slits, and can be continually changed by the contraction of the disk, the bursæ serve as respiratory organs in addition to their primary function of discharging the genital products. The minute tentacles of the arms also aid in respiration.

DEVELOPMENT

In most species of ophiurans the genital products are simply discharged into the sea water, and abandoned to their fate. In the little *Amphipholis squamata*, however, the eggs are retained within the body of the parent until the embryonic development is completed and the little ophiuran has reached the form of the adult. Moreover, in this case, a nourishing exudation is said to be poured out upon the eggs, which supplements the store of food contained in the yolk.

This species is likewise hermaphrodite, both ovary and spermary being attached to each of the genital bursæ. After the young have reached the adult form, the whole disk of the parent, with the exception of the mouth frame, may sometimes be thrown off to allow the young to escape, and then be regenerated. The breeding habits and development of this species have been studied by Fewkes* and others.

In the majority of ophiurans, the sexes are separate, the eggs being thrown out into the water, and fertilization takes place whenever suitable sperm cells are in the vicinity.

The eggs segment by a regular cleavage, as in the starfish and sea-urchin. As in these two groups, the embryos quickly develop into free-swimming larvæ, which in this class are called *plutei,* and which often feed for several weeks at the surface of

---

* *Bull. Museum Comparative Zoology,* vol. xiii, 1887.

the sea. As shown in Fig. 11, the plutei in no way resemble the adult, but are somewhat pyramidal in form, and often have

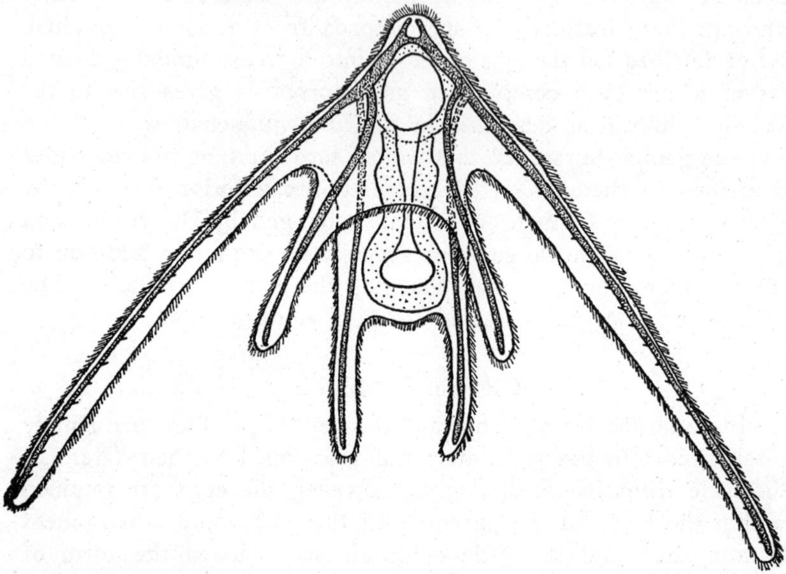

FIG. 11. *Ophiopholis aculeata.* Free-swimming larva, with long, ciliated arms. (Modified from Fewkes.)

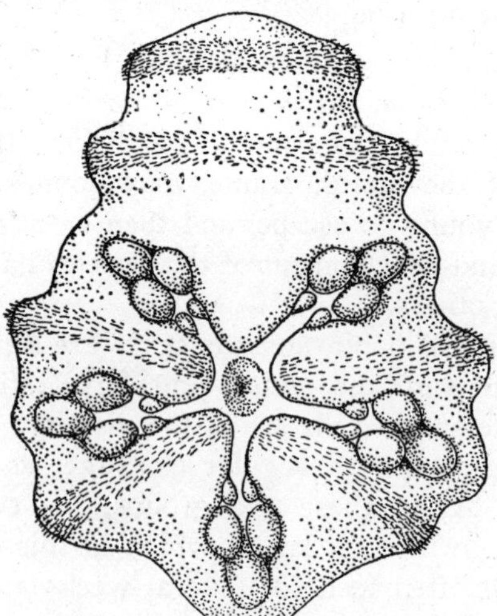

FIG. 12. *Ophiura brevispina.* Free-swimming larva preparatory to transformation to the adult form. (After Grave.)

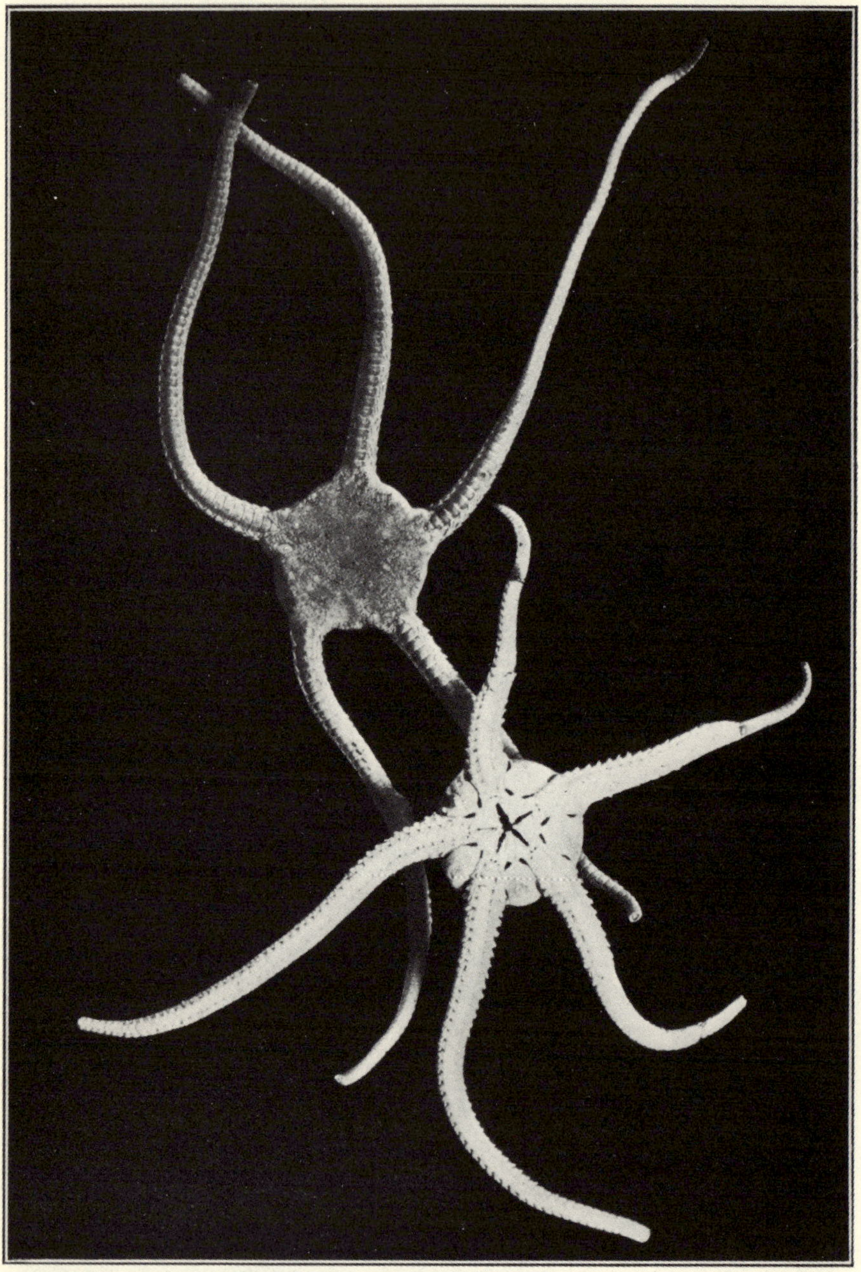

PLATE XV. *Ophiura brevispina.* (Twice natural size.)

Oral and aboral surfaces of dried specimens. The fragile tips of several of the rays have been broken off, and the lower figure shows one ray in process of regeneration.

extremely long, ciliated arms. Fig. 12 shows the larva of another species, in which there are four bands of cilia in place of the long arms. The metamorphosis is accomplished in a manner similar to that described for the starfish.

KEY TO SPECIES

The ophiurans of the world are commonly divided into three orders, distinguished by the character of the vertebræ. Only one of these orders, the **Zygophiuræ**, with interlocking vertebræ, is represented in Long Island Sound.

1. Spines on sides of arms short, and rather closely pressed against the arms .... **Ophiura brevispina**
   Spines projecting from sides of arms at a marked angle. . 2
2. Large median plates on upper side of arms surrounded by a series of small plates; 5 or 6 spines on each side of each joint of arm .... **Ophiopholis aculeata**
   Median plates on upper side of arms directly in contact; that is, not surrounded by small plates............... 3
3. With 3 tooth-like projections (oral papillæ) along each edge of each jaw; length of arms less than 10 times the diameter of disk; radial shields in close contact **Amphipholis squamata**
   With 4 or 5 oral papillæ on each edge of each jaw; length of arms more than 10 times the diameter of disk; radial shields separated 4
4. Oral papillæ 4........................ **Amphioplus abditus**
   Oral papillæ 5.................... **Amphioplus macilentus**

Ophiura brevispina Say, var. **olivacea**

Green Ophiuran

Plate I, fig. 5; Plates XIV, XV.

This species, which was described by Say in 1825, is known in the earlier writings of Lyman and Verrill under the name of *Ophiura olivacea.* It has been found in shallow water at numerous localities along the Atlantic coast from Cape Cod to Brazil, and at Bermuda. It is sometimes found, however, at a depth of over a hundred fathoms, especially in the warmer regions.

The form occurring north of Florida is now usually considered a distinct variety, *olivacea.*

This ophiuran occurs at numerous localities in Long Island Sound, living in water but a few feet deep at low tide, and is partial to protected situations and tide pools. It is often abundant on sandy shores among the roots of eelgrass. In such places the bottom is often covered with a mass of dead grasses and algæ, and forms the home of numerous other small animals, such as worms, mollusks, and crustaceans. Here the ophiuran hides during the daytime and is seldom seen unless disturbed. The snake-like movements of the arms may then attract attention, but the color of the animal when resting agrees most closely with that of the surrounding objects, and the appearance of the arms is quite similar to that of fragments of the stems of the eelgrass.

The color is commonly greenish or brownish but is sometimes nearly black. Olive-green individuals are perhaps the most common. The disk is often mottled and the arms banded with light and darker shades, but the colors are always dull and inconspicuous.

The disk is usually 10 to 15 mm. in diameter, and the arms 40 to 60 mm. in length. The arms are rounded and moderately slender, as indicated on Plates I and XV. Each segment of the arm has seven or eight short spines on each side, and these lie closely appressed to the arm, so that they are quite inconspicuous, as shown on Plates XIV and XV.

A most interesting account of the habits and development of this species has been published by Grave.*

Ophiopholis aculeata (Linnæus)

Daisy Brittle Star

Plates XII, XIII, XVI.

This is without question the most beautiful and most conspicuous of all our native ophiurans, but unfortunately it is but rarely found in Connecticut waters, and there occurs only in the deeper areas toward the eastern portion of Long Island Sound. North of Cape Cod the species is very abundant and is there

* Ophiura brevispina. *Mem. Nat. Acad. Sciences,* vol. viii, 1899.

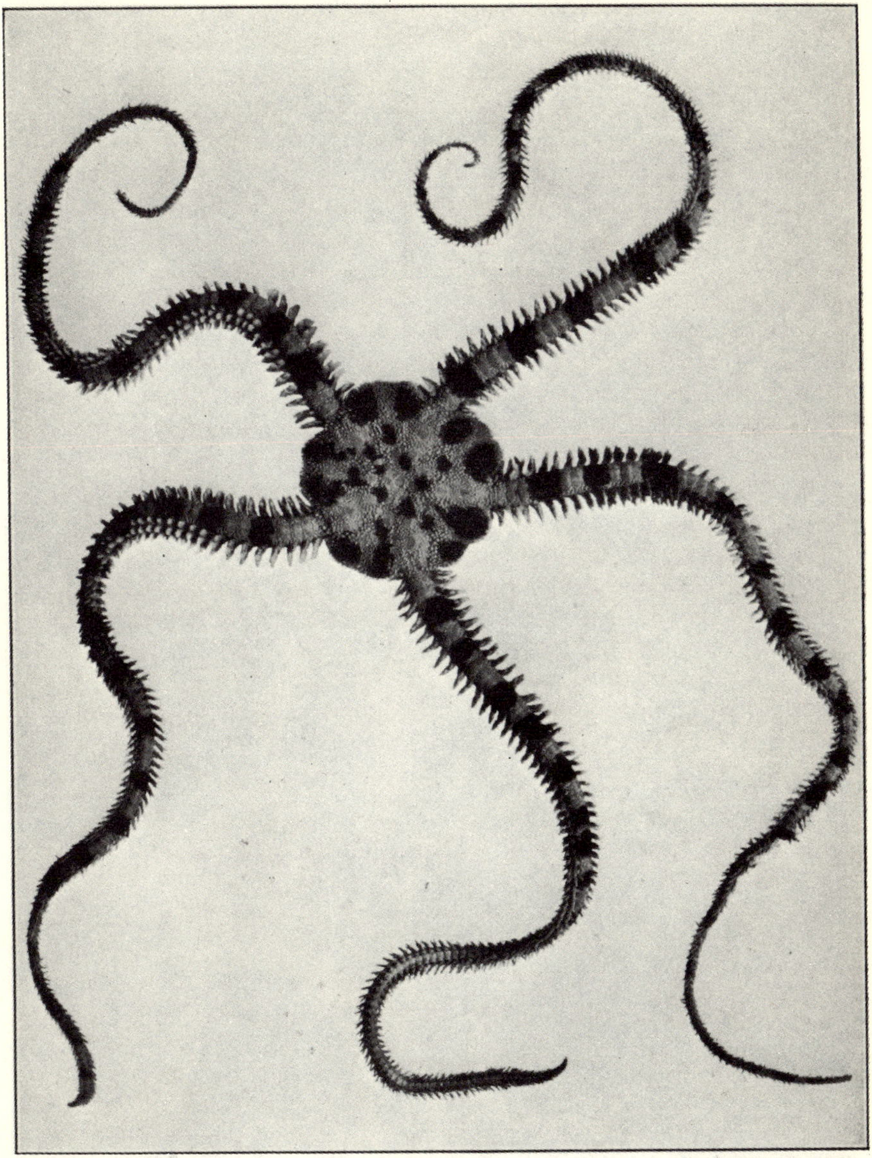

PLATE XVI. Daisy Brittle-star, *Ophiopholis aculeata.* (One and one-half times natural size.)

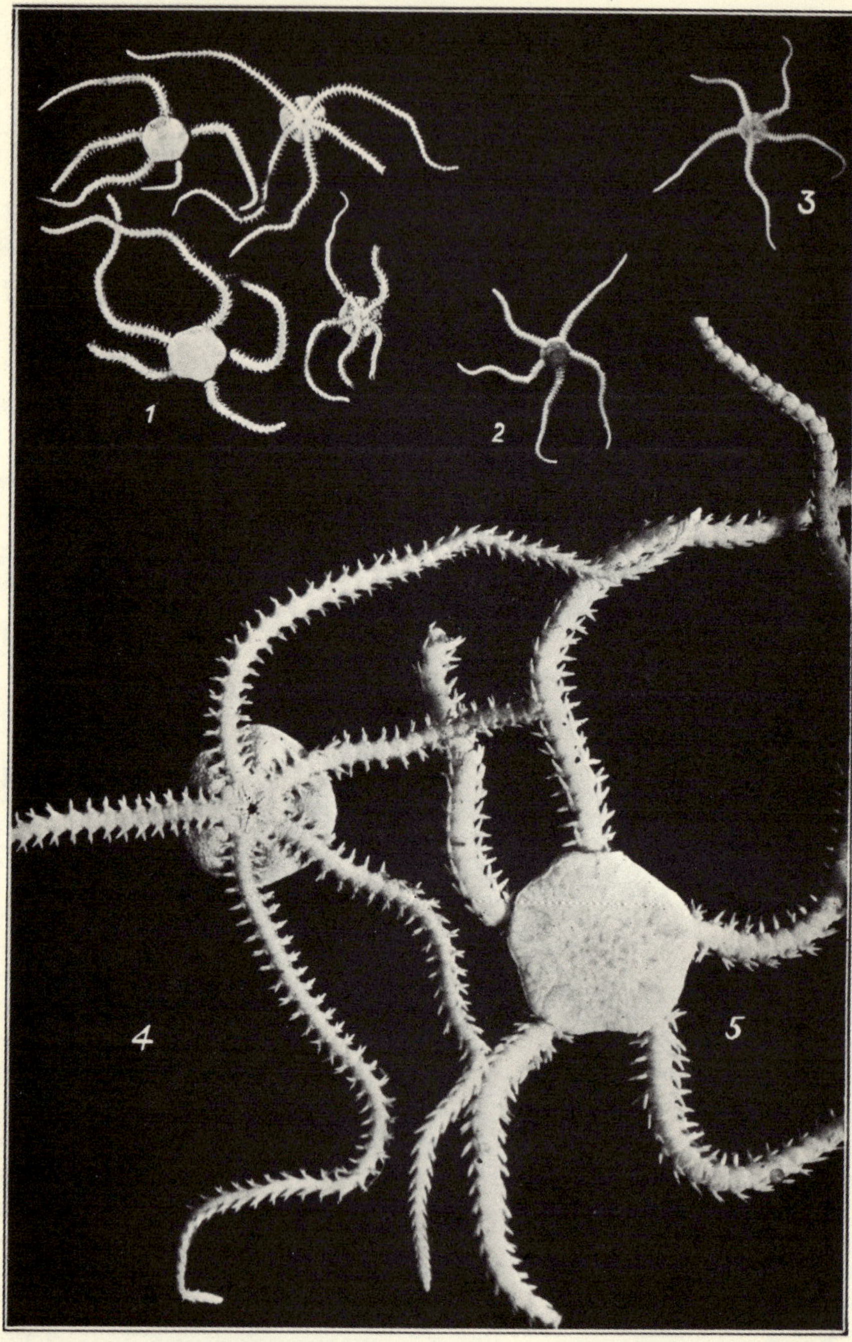

PLATE XVII. *Amphipholis squamata.* (Fig. 1, one and one-half times natural size. Figs. 2 and 3, natural size. Figs. 4 and 5, ten times natural size.)

commonly known as the "daisy brittle star." It extends northward to the Arctic Ocean, and on the European coast reaches southward as far as the English Channel. It is larger than any other of our species, the disk often being 15 to 20 mm. in diameter, and each of the arms 50 to 80 mm. in length.

The colors are exceedingly variable, not only in hue but also in the arrangement of the patterns. There are all shades of red, purple, brown, yellow, and green. The disk is spotted and blotched in some regular pattern arranged either radially or concentrically or both, while the arms are either banded or striped lengthwise (Plate XII.) The bright colors are limited to the aboral surface, the under side of disk and arms being commonly gray or yellowish, except that the dark bands on the arms often extend more or less completely over the oral surface. Of a hundred individuals taken at random scarcely two will agree closely in both shade and arrangement of markings, so that the individuality of the representatives of this species of ophiuran is more conspicuous than in almost any other invertebrate.

Amphipholis squamata (Delle Chiaje)

Plate XVII.

This is the smallest species of ophiuran recorded on our coast, and it is also the most abundant and the most widely distributed. On the Atlantic coast of America it occurs from the Arctic Ocean to the coast of New Jersey, and it is also found on the northern coasts of Europe and in the Mediterranean Sea. In America the species is often referred to under the name of *Amphipholis elegans* or *Amphiura elegans,* and in Europe is sometimes called *Amphiura squamata.*

Explanation of Plate XVII. *Amphipholis squamata.*

Fig. 1. Four specimens showing both oral and aboral surfaces. The tips of several of the very fragile rays have been broken off. One and one-half times natural size.

Fig. 2. Aboral surface. Natural size.

Fig. 3. Oral surface. Natural size.

Fig. 4. Oral surface. Enlarged ten times.

Fig. 5. Aboral surface, showing the smooth scale-like plates which cover the disk. The narrow radial shields are in close contact with each other. Each segment of the arms has three arm spines on each side. Enlarged ten times.

(Figs 2 and 3 from photographs loaned by the U. S. Fish Commission, with permission of Dr. H. L. Clark.)

In Long Island Sound this ophiuran occurs on the eelgrass in New Haven harbor, and has been collected at several other localities in shallow water, hiding in dead mollusk shells and in the interstices of stones.

The diameter of the disk is 3 to 5 mm., and the length of each of the five slender arms is from 10 to 20 mm.

The color is gray or pale brownish, often with numerous lighter or whitish spots. The oral surface is gray with a yellowish tinge, usually with a whitish spot at the base of each arm. In very young specimens the disk is orange.

The disk is flattened and nearly circular (Plate XVII), and is covered with small scales. The arms have three spines on each side of each segment, and these stand out at nearly right angles to the axis of the arm, as shown in the figure.

A detailed description of the arrangement of scales on disk and arms is to be found in Clark's Report on the Echinoderms of the Woods Hole region.*

In this species the eggs mature in midsummer and are retained in the genital bursæ after fertilization. They there develop into young ophiurans, which are then discharged from the body as tiny brittle stars, of form and structure like the parent. They do not undergo the metamorphosis characteristic of the other species, for none of our other species is known to be viviparous.

Amphioplus abditus Verrill

Plate XVIII; Plate XIX, fig. 1.

This locally distributed species was for many years known only from Long Island Sound. Quite recently, however, a single specimen has been recorded from the harbor at Woods Hole, Massachusetts,† and one specimen was collected by the writer some years ago at the same locality.

The animal lives buried in the soft mud, and when exposed quickly buries itself again. Verrill‡ states that it naturally lies with one or more of the very slender arms projecting above the surface of the mud. "On this account it is seldom dredged

* *Bulletin U. S. Fish Commission* for 1902, p. 560.

† H. L. Clark, *Science*, vol. xxvii, p. 147, 1908.

‡ Invertebrate Animals of Vineyard Sound, p. 139.

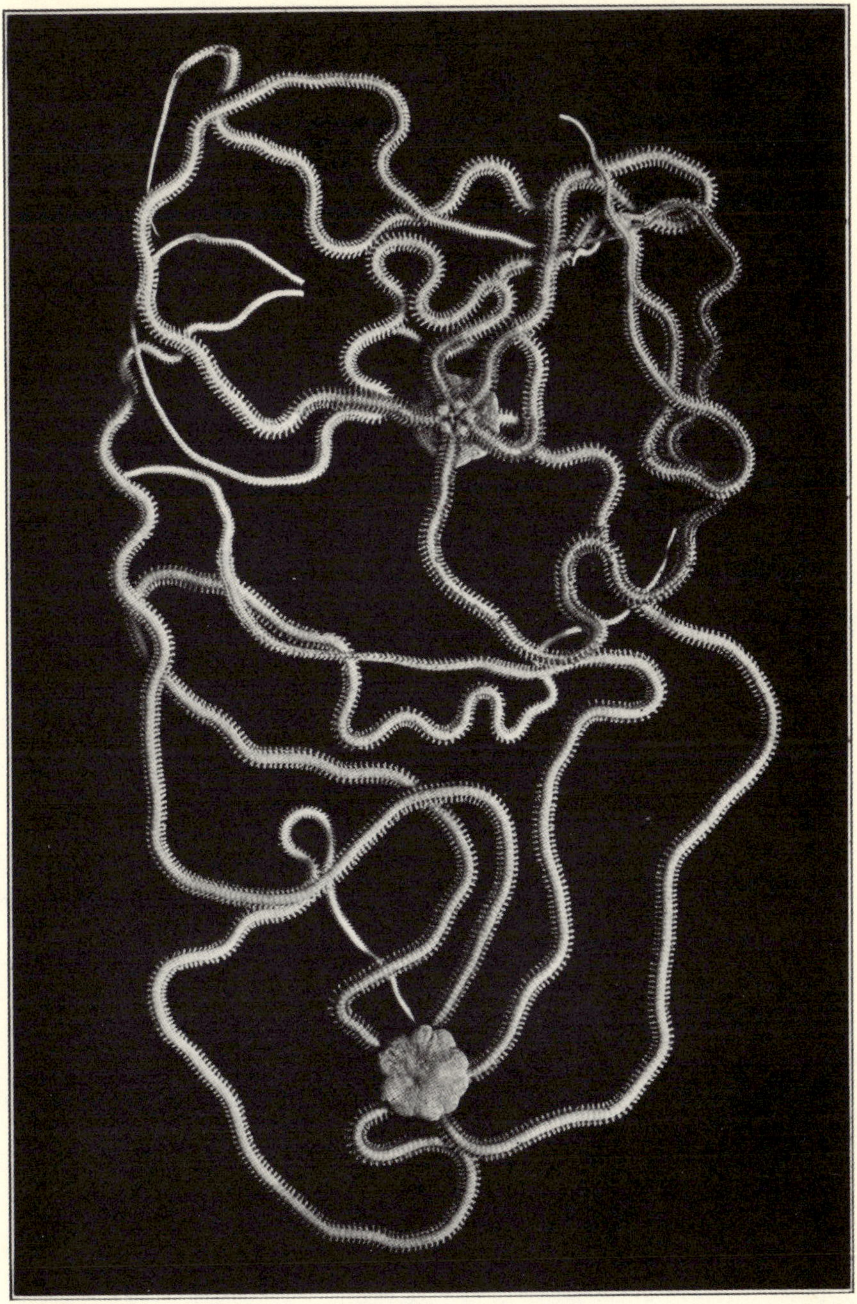

PLATE XVIII. *Amphioplus abditus.* (Slightly enlarged.)
Two individuals, showing both oral and aboral surfaces of disk, and the extremely slender rays.

entire; the projecting arms are usually cut off by the dredge, and the animal escapes; and as it has the power of restoring lost arms, this is only a temporary inconvenience. The same thing probably happens when a voracious fish seizes one of the arms."

The specimens from Woods Hole were collected at extreme low-water mark. In Long Island Sound the species has been dredged off New Haven and at Thimble Islands at depths of from three to eight fathoms, and many specimens have been collected in soft mud, near low-water mark, at Noank, Connecticut, and others at Peconic Bay, Long Island.

The disk of mature individuals measures from 6 to 11 mm. in diameter. The arms are very slender and extremely flexible. They are usually from twelve to sixteen times as long as the diameter of the disk, and sometimes measure as much as 180 mm. in length.

In addition to the extremely long and slender arms (Plate XVIII, Plate XIX, fig. 1), the species may be distinguished by other anatomical peculiarities. Careful examination of the jaws will show the presence of four tooth-like processes, oral papillæ, on each edge. On the sides of each segment of the arm is a series of three projecting spines, of which the middle one is stouter, flatter, and less pointed than the other two.

On the aboral side of the disk at the base of each arm is a pair of narrow plates, called the radial shields. In the present species these are narrow, curved, and parallel or only slightly divergent, and not closely in contact. They are often partly covered by the encroachment of the small scales which cover the disk. The scales of the disk are without granulations or spines.

The color of the living animal varies considerably in different specimens. Some are brown on the aboral surface of the disk, with a central area of dark brown bordered by pale gray. A radiating band of dark brown likewise bordered by gray extends outward to each interradial margin. The radial shields are yellowish brown, while the narrow space between them is blue. There are additional symmetrically placed brown spots arranged in series on the disk. The arms are brown or dull greenish, often with lighter bands. The specimen from Woods Hole studied by Clark was uniformly gray.

The animals have remarkable powers of regeneration, for they have the habit of throwing off the disk when violently disturbed, leaving only the mouth frame and adhering organs attached to the arms. Under favorable conditions they are able to regenerate the missing disk, and specimens have been collected in which such regeneration was actually in progress.

Full details of the anatomical peculiarities are given in Verrill's original description of the species.*

The species is also fully described in Verrill and Smith's Invertebrate Animals of Vineyard Sound, page 426, although the description there given has been revised in certain details in a later paper by Verrill.†

Amphioplus macilentus Verrill

Plate XIX, fig. 2.

This species is closely related to the preceding, but is distinguished by having five oral papillæ in each row, and the spines at the sides of the arms are all slender.

Although this species has not yet been recorded from Long Island Sound, it has been collected at various localities so near the eastern entrance to the Sound as to render it highly probable that further investigation will reveal its presence in Connecticut waters. It has been dredged by the United States Fish Commission in muddy localities at numerous stations off the southeastern coast of New England, and in certain places occurs in large numbers as far south as Cape Hatteras.

Full details of the anatomical peculiarities are given in Verrill's original description of the species.‡

The individuals belonging to this species are small, with very long and slender arms (Plate XIX, fig. 2). The disk is

* *Am. Journ. Sci.*, 3rd series, vol. ii, p. 132, 1871.

† North American Ophiuroidea. *Trans. Conn. Acad.*, vol. x, p. 314.

‡ *Am. Journ. Sci.*, 3rd series, vol. xxiii, p. 142, 1882.

EXPLANATION OF PLATE XIX.

Fig. 1. *Amphioplus abditus.* Oral surface of disk and portions of rays. One of the extremely slender rays is in process of regeneration. Three times natural size.

Fig. 2. *Amphioplus macilentus.* Oral and aboral surfaces, showing the extreme slenderness of the rays. Twice natural size.

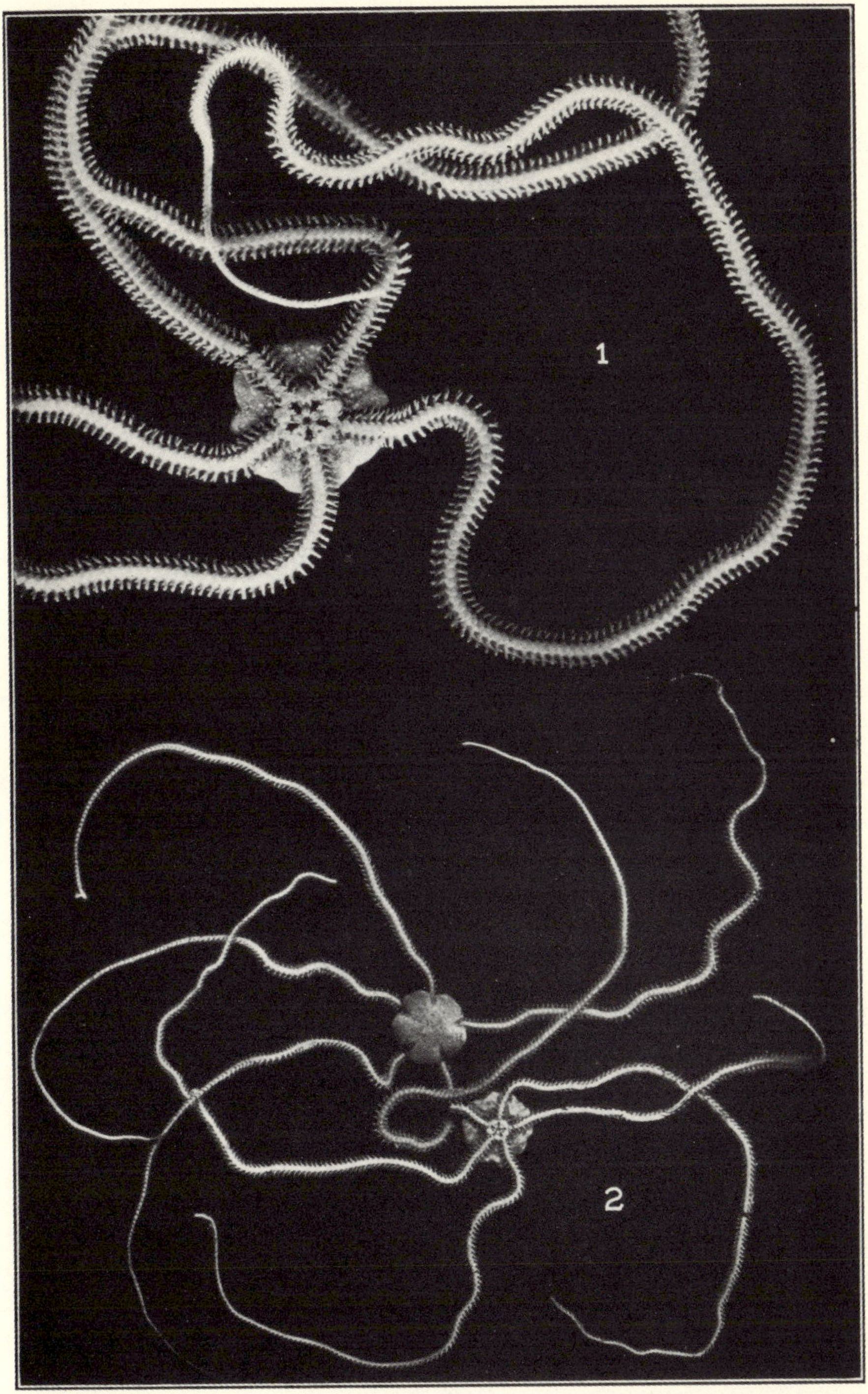

PLATE XIX. Fig. 1, *Amphioplus abditus.* (Three times natural size.)
Fig. 2. *Amphioplus macilentus.* (Twice natural size.)

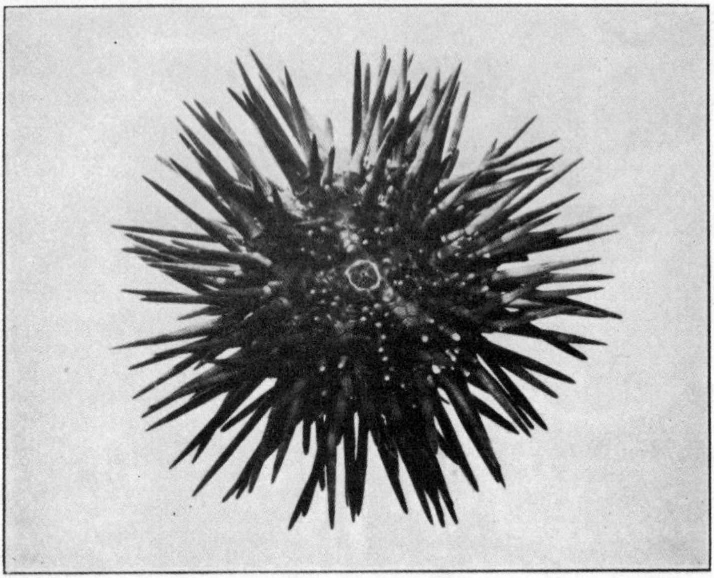

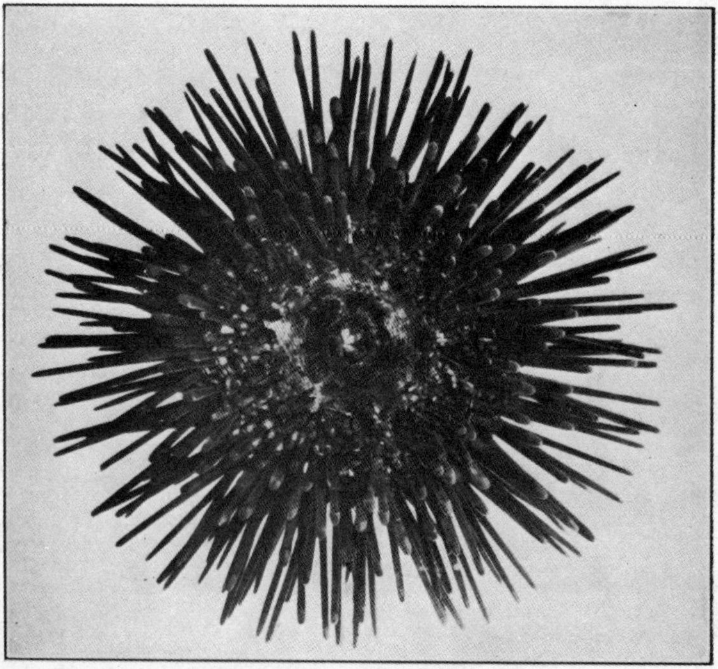

PLATE XX. Purple Sea-urchin, *Arbacia punctulata*. (Natural size.)

nearly circular in outline in the living animal, but often becomes pentagonal when preserved in alcohol or dried. The aboral surface of the disk is covered with numerous, small, naked, imbricated scales, forming a rather distinct rosette at the center. The radial shields are long and narrow, with their outer ends in contact, and their inner ends separated by a narrow wedge of small scales. The oral surface is covered with more minute scales than those of the aboral surface. The oral shields are shield-shaped, rather longer than wide.

The five oral papillæ are unequal in size, but all are obtusely rounded. The inner one is the largest and stoutest, while the next is the smallest and most acute; the next two are flattened, while the outermost is small and rounded.

The under arm plates are shield-shaped and longer than broad. Near the base of the arms they are in contact, but farther out they are separated by the side arm plates. Tentacle scales two. Arm spines three, nearly equal, about as long as the joints.

Color light gray. Diameter of disk about 4 mm. Length of the extremely slender arms about 60 mm.

CLASS III. ECHINOIDEA

This class includes the animals popularly known as sea-urchins, sand-dollars, cake-urchins, disk-urchins, sea-cakes, and sand-cakes. In the sea-urchins the body is nearly hemispherical, while the others are flattened and discoid. Spines thickly cover the body on all sides, except sometimes at the poles. In one of our species (*Mellita pentapora*) the disk is pierced by five oval perforations.

STRUCTURE

If the purple sea-urchin (*Arbacia punctulata*) be taken as an example of this group, the body is seen to be enclosed in a

---

EXPLANATION OF PLATE XX. Purple Sea-urchin, *Arbacia punctulata.* (Natural size.)

The larger specimen (lower figure) shows the oral surface, with the five sharp white teeth in the center; while the smaller specimen (upper figure) shows the aboral surface, in the center of which is the periproct, surrounded by a broad area free from spines. (After Clark.)

firm, hemispherical shell made up of 20 meridional rows of closely fitting plates, and with a wide circular aperture at one pole (Plate XXV). In a living specimen of this same species (Plate XX), the shell, or test, is so thickly covered with long stiff spines as to obscure its surface completely except at the poles.

At one pole of the body may be seen the five sharp, calcareous teeth surrounding the mouth. This is therefore the oral, or actinal pole of the body; the opposite pole being aboral, or abactinal. Around the mouth is a broad area devoid of large spines, called the peristone. It consists of a strong membranous wall with an abundant musculature, enabling the mouth to be opened and closed. The peristome is provided with numerous slender pedicellariæ, which differ from those of the starfish in having three blades, which form the angles of a triangle. Similar pedicellariæ are scattered over the body. The circular lip surrounding the teeth can close the mouth very tightly (Plate XX).

At the aboral pole there is likewise a considerable area (Plate XX) devoid of spines, in the center of which is a cluster of movable plates (Fig. 13) constituting the periproct. The intestine opens in the center of the periproct. In the species under consideration the periproct consists of four triangular plates, while in the other native species of sea-urchin (*Strongylocentrotus dröbachiensis*) there are many minute plates (Plate XXI). The spines which cover the surface of the test vary somewhat in shape, but all are cylindrical, solid, and longitudinally fluted. These spines not only constitute a very formidable defensive covering, but are used in locomotion. Some of the spines nearest the aboral pole are sharply pointed, while others are blunt; below the equator of the test the tips of the long spines are somewhat flattened and rounded, while those nearest the peristome are short, and spatula-like, with flattened, rounded tips (Plate XX).

The base of each spine has a rounded socket which fits closely against a corresponding hemispherical tubercle on one of the plates of the test, allowing a great range of movement. Closely fitting this ball-and-socket joint is a wreath of muscular fibers attached both to test and spine. These muscles are able to move the spine in all directions. Beneath these "moving" muscles is a wreath of specialized muscle fibers whose function is, when stimulated,

PLATE XXI. Skeleton of Sea-urchin, *Strongylocentrotus dröbachiensis*. (Natural size.)

to set the spine stiffly in one direction. As long as the action of this muscle lasts the spine can only be moved by tearing the muscle.

The test denuded of spines (Plate XXV) shows clearly the regularity with which the spines are arranged, each tubercle indicating the position of a spine. In *Strongylocentrotus* there are numerous smaller secondary spines between the larger ones, as shown by the smaller tubercles (Plate XXI). In this genus there are several times as many spines as in *Arbacia.*

Both the tube-feet and the spines assist in locomotion, as described below.

An examination of a living specimen shows five pairs of meridional groups of very slender tube-feet, which are capable of expanding to great length. The extremity of each tube-foot on the oral surface has a sucker-like expansion, kept open by a perforated calcareous plate. The position of the tube-feet is indicated in the denuded test (Plate XXV) by the ten series of plates with minute perforations. These ten series of openings correspond with the groups of tube-feet seen in the living animal, each pair of openings representing the position of a single tube-foot.

The series of ten rows of plates which bear the tube-feet constitute the five ambulacral, or radial areas, there being two rows of plates to each. Between the ambulacral areas are five interambulacral, or interradial, areas of about equal extent. Each of these five areas likewise consists of two meridional rows of plates (Fig. 13).

The ambulacral areas may thus be seen to represent the oral surfaces of the arms of the starfish, bent upward in a semicircle, while the interambulacral areas correspond to the sides of the

---

EXPLANATION OF PLATE XXI. Skeleton of Sea-urchin, *Strongylocentrotus dröbachiensis.* (Natural size.)

Figs. 1 to 3. Aboral, lateral, and oral surfaces with spines removed, showing the regularly arranged rows of larger and smaller tubercles which in life bear the primary and secondary spines respectively. Scattered among the larger tubercles are numerous minute tubercles which bear the miliary spines. The oblique rows of minute dots arranged in pairs represent the pores for the tube-feet, and indicate the position of the five double rows of plates of the ambulacral areas. Of about equal extent are the five interambulacral areas. In Fig. 1 the centrally placed periproct, consisting of many minute plates, is surrounded by a circle of larger ocular and genital plates. In Fig. 3 the peristome and jaws have been removed.

arms, firmly fitted together. Thus the entire aboral surface of both disk and arms of the starfish can be represented in the sea-urchin only by the few plates of the periproct.

Each ambulacral area ends near the periproct in a single plate called the ocular plate (Fig. 13), perforated by a pair of minute openings from which a tentacle protrudes, as at the end of the arm of the starfish. The interambulacral areas likewise terminate in single large plates called the genital plates, because they bear the five openings of the genital glands (Plate XXV). One of these genital plates is larger than the others, and is perforated by numerous minute pores which lead to the water-vascular system. This plate is therefore termed the madreporic plate (Fig. 13).

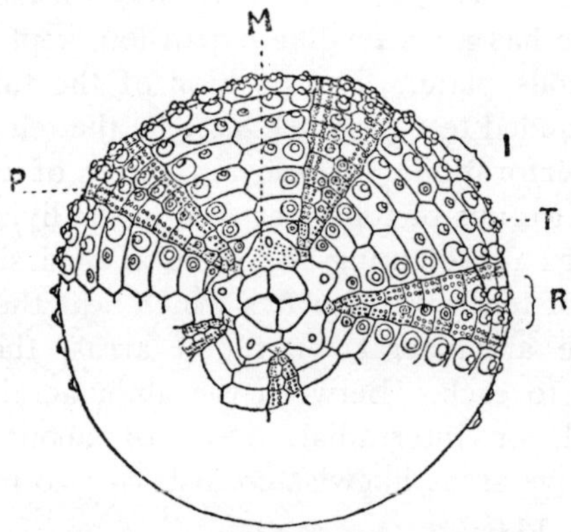

FIG. 13. *Arbacia punctulata.* Outline of plates of aboral surface, showing the four plates of the periproct, surrounded by the five ocular and five genital plates at the ends of the ambulacral (*R*) and interambulacral (*I*) zones, respectively; the genital plate marked *M* is the madreporic plate; *T*, tubercles for attachment of spines; *P*, pores for the tube-feet.

At the margin of the peristome are five pairs of much branched processes, having a respiratory function, and termed the dermal branchiæ, or gills.

Just inside the test at the border of the peristome, and resting upon the bases of the ambulacral areas, are five pairs of calcareous

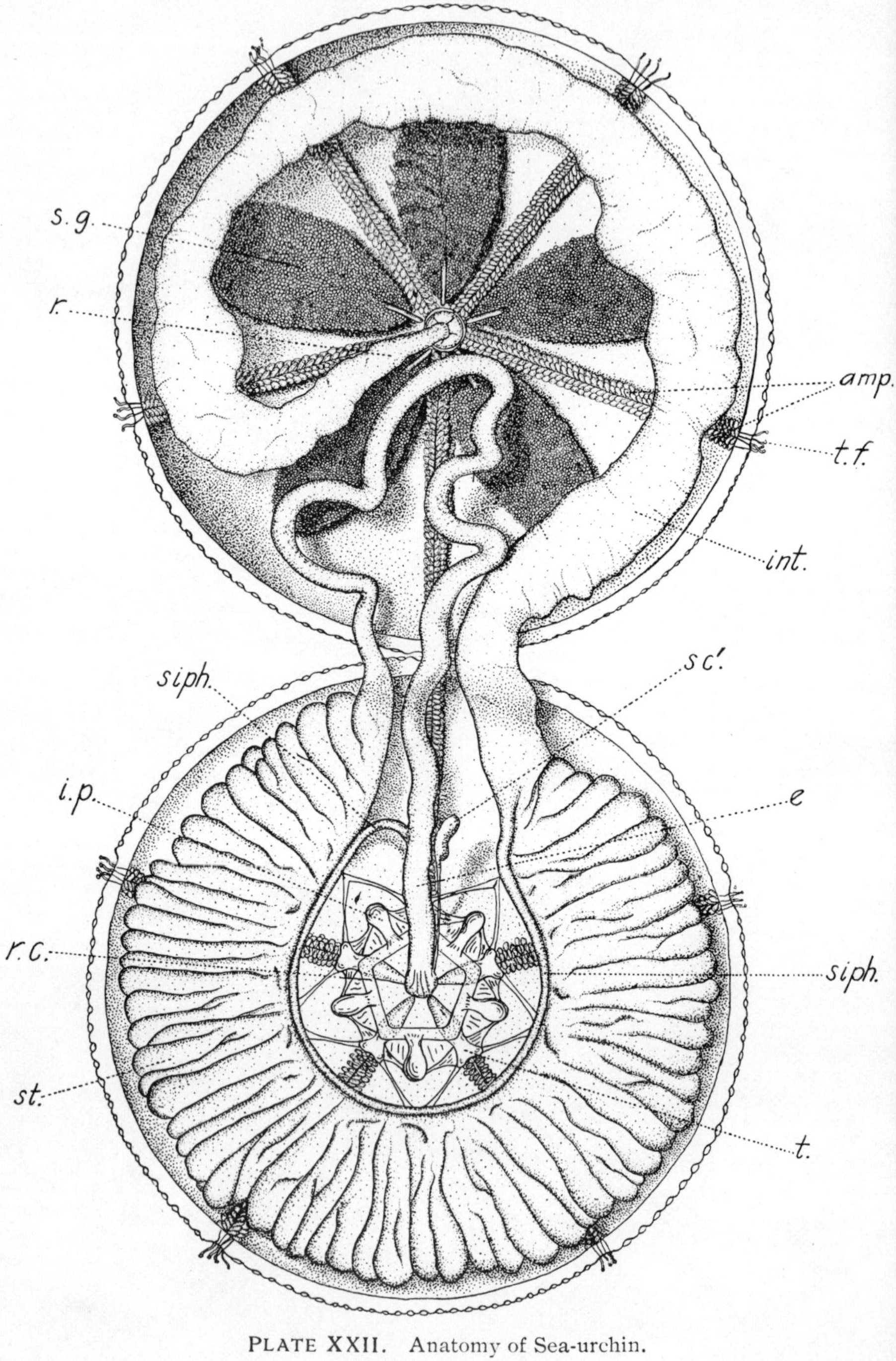

PLATE XXII. Anatomy of Sea-urchin.

pillars, the auricles, to which are attached the powerful muscles for protruding and retracting the jaws. In *Strongylocentrotus* the pillars of each pair are connected by a bridge.

Internal Anatomy.— If the test is cut through at the equatorial region, and the two portions laid side by side without injuring the alimentary canal, the internal organs of the sea-urchin are easily studied. The more prominent organs to be seen in such a dissection are shown on Plate XXII.

Digestive System.— In the sea-urchin, the mouth, situated in the center of the oral surface, is provided with five sharp calcareous teeth which meet closely in the axis of the body and close the mouth completely. These teeth are set in jaws which are moved by a complicated set of muscles attached to a calcareous frame-work; this, with the attached teeth, is often called the lantern. When the jaws are separated the mouth presents a wide opening leading into the esophagus. The latter is short, and at its end divides into two branches, of which one is coiled irregularly for some distance and opens into the wide, saccular stomach, while the other forms the siphon (Plate XXII).

The stomach is broad and flat; its walls are irregularly constricted to form numerous sac-like pouches which greatly increase the surface area. It makes a nearly complete circle at the equator of the body and then bends sharply upward and passes into the narrower intestine. The intestine is also flat, and at its beginning bends directly backwards and encircles the body. Its course lies parallel with the stomach but it leads in the opposite direction.

Explanation of Plate XXII. Anatomy of Sea-urchin. *Arbacia punctulata,* showing the arrangement of the internal organs. The test has been cut horizontally and the two parts laid side by side without cutting the alimentary canal; the upper circle representing the aboral portion of the test and the lower circle the oral part. At the center of the lower section the mouth opens into the esophagus (*e*) from which the spihon (*siph*) and stomach (*st*) branch out. The intestine (*int*) in the upper section lies parallel with the stomach in its natural position, but its course is in the opposite direction. The rectum (*r*) opens through the anus at the center of the upper section. The five sexual glands (*s. g*) open by narrow ducts on the aboral surface near the anus. The five teeth (*t*) lie at the five angles of the mouth. The ring canal (*r. c*) with the five interradial pouches (*i. p*) are shown. Beside the esophagus lies the delicate madreporic canal with its dorsal ampulla (*s. c'*) although the connection of the latter with the aboral surface has necessarily been torn away. The rows of ampullæ (*amp*) and tube-feet (*t. f*) connected with the radial canals are shown in both sections.

After making a nearly complete circle, the intestine passes gradually into the rectum, which bends obliquely upward and opens to the exterior in the center of the aboral surface.

The siphon makes a short cut between the esophagus and the posterior end of the stomach. It consists of a narrow tube imbedded in the inner wall of the stomach, and is lined throughout with a strongly ciliated epithelium, which presumably draws a stream of water from the esophagus to the intestine. This water doubtless aids in respiration, and may also serve to wash indigestible food materials out of the alimentary canal.

The food of the sea-urchin consists of seaweeds and various small animals. Some species devour such dead animal matter as can be obtained, individuals sometimes collecting in numbers on the body of a dead fish. The organic matter on the surface of the sea bottom is also used for food, while certain species are said to capture and devour large crustaceans. The sharp chisel-like teeth would seem to fit the creature for devouring almost any kind of plant or animal matter which by its slow method of locomotion it is able to secure.

Water-vascular System.— From the madreporic plate, the madreporic canal leads directly down to the circular vessel, which is situated immediately above the jaws and surrounds the base of the esophagus. Connected with the circular vessel are five interradial pouches, or Polian vesicles (Plate XXII).

Five radial vessels, similar to those of the starfish, leave the circular vessel, pass to the borders of the peristome and thence along the inside of the test directly beneath the ambulacral areas, and end in the terminal tentacle of the ocular plate. The radial vessels supply the tube-feet as in the starfish, each tube-foot connecting with an ampulla inside the wall of the test (Plate XXII). Two perforations of the plates of the test are provided for each tube-foot, so that the contained fluid may enter the foot by one canal and return to the ampulla by the other. Internal cilia keep up this circulation of fluid, and by this means the fluid in the tube-feet is continually changing. An exchange of gases can take place between the fluid contained in the tube-feet and the sea water, and a supplementary respiration is accomplished thereby. In *Arbacia* only the tube-feet of the lower part of the body are provided with suckers and used for locomotion, those of the upper part being specially adapted for respiration, while in *Strongylocentrotus* all the tube-feet have suckers.

Nervous System.— As in the starfish, there is a main nerve ring surrounding the mouth, with five radial nerves which accompany the radial vessels to the base of the terminal tentacle. Each tube-foot and all the spines and pedicellariæ are supplied with branches of these radial nerves.

The tube-feet are not only used for locomotion and respiration, but are highly sensitive to various kinds of stimuli. Specialized tube-feet on the peristome enable the animal to taste. The pedicellariæ are in many species modified to perform a variety of functions, some having specialized sense organs and poison glands. Minute organs scattered over the body in many species, and known as sphæridia, are thought by some to act as balancing organs.

Reproductive System.— Attached by strong mesenteries to the upper wall of the test are the five sexual glands. The genital duct leading from each of these glands opens by a minute pore in the corresponding genital plate adjacent to the periproct. When fully developed the sexual glands become enormous masses of tiny follicles (Plate XXII).

The sexes are separate, but are not distinguishable externally. The females produce many thousand eggs, which are discharged into the water when mature.

Comparison of Arbacia with other Genera.— A comparison of our three other representatives of this class with the description given above for *Arbacia* offers many points of interest.

The green sea-urchin agrees with the purple sea-urchin in having a nearly hemispherical test, with the mouth placed symmetrically at one pole and the anal opening at the other. Both species therefore belong to the order Regularia, or radially symmetrical urchins. Both the sand-dollar and the key-hole urchin, on the other hand, have flat, discoid bodies, nearly circular in outline but with one or more slight indentations; and the key-hole urchin has five narrow oval perforations (Plate XXVIII).

In the sand-dollar and the key-hole urchin, the mouth occupies the usual position in the center of the oral surface; but in the sand-dollar the anus has been shifted to the edge of the disk (Plate XXIII), and in the key-hole urchin it is very near the mouth, on the proximal border of the middle lunule (Plate XXIV, fig. 1). In neither of these forms, therefore, is the body radially symmetrical, and they belong to the order Clypeastroidea, or disk-urchins, as they may be called.

Both these species of clypeastroids have strong jaws, with fine, sharp, calcareous teeth, but the jaws (Plate XXIV) are less complicated than in the order Regularia. The peristome is very small, so that the jaws cannot be opened so widely as in the regular forms. By means of the teeth both these species pick up particles of sand and mud, with the diatoms and other minute organisms adhering. These particles are swallowed in vast numbers, and the organic matter digested out, after which the sand is passed out of the anal opening.

Both of our species of disk-urchins are plainly bilaterally symmetrical, the mouth and anus lying in the median antero-posterior plane. This bilaterality is shown also in the outline of the body. As illustrated in Plate XXIV, the anus in *Echinarachnius* lies in an indentation at the posterior end of the disk, and there are two other slight indentations on the edge of the disk in the adjacent ambulacral areas. In *Mellita* (Plate XXVIII) the manifestation of bilateral symmetry is carried still farther, for here there are five symmetrically placed lunules, one being in the posterior interradius, while the others occupy the two adjacent radii on each side. These arise in the young urchins as notches in the edges of the test, and gradually deepen and close over as the animal increases in size. Furthermore in both species the disk is broader posteriorly, and narrower in front.

Although the anal opening has moved from its primitive position at the aboral pole, the ocular and genital plates have retained the same position as in the regular sea-urchins. They

---

EXPLANATION OF PLATE XXIII. Skeleton of Disk-urchin, *Echinarachnius parma.* (Natural size.)

Aboral and oral surfaces with spines removed. The upper figure shows the ambulacral areas ("petals") and the four genital pores near the center of the disk. The lower figure shows the mouth and teeth, and the closely fitted plates. The indentation on the lower median border indicates the position of the anus.

(Photographs loaned by the U. S. Fish Commission, with permission of Dr. H. L. Clark.)

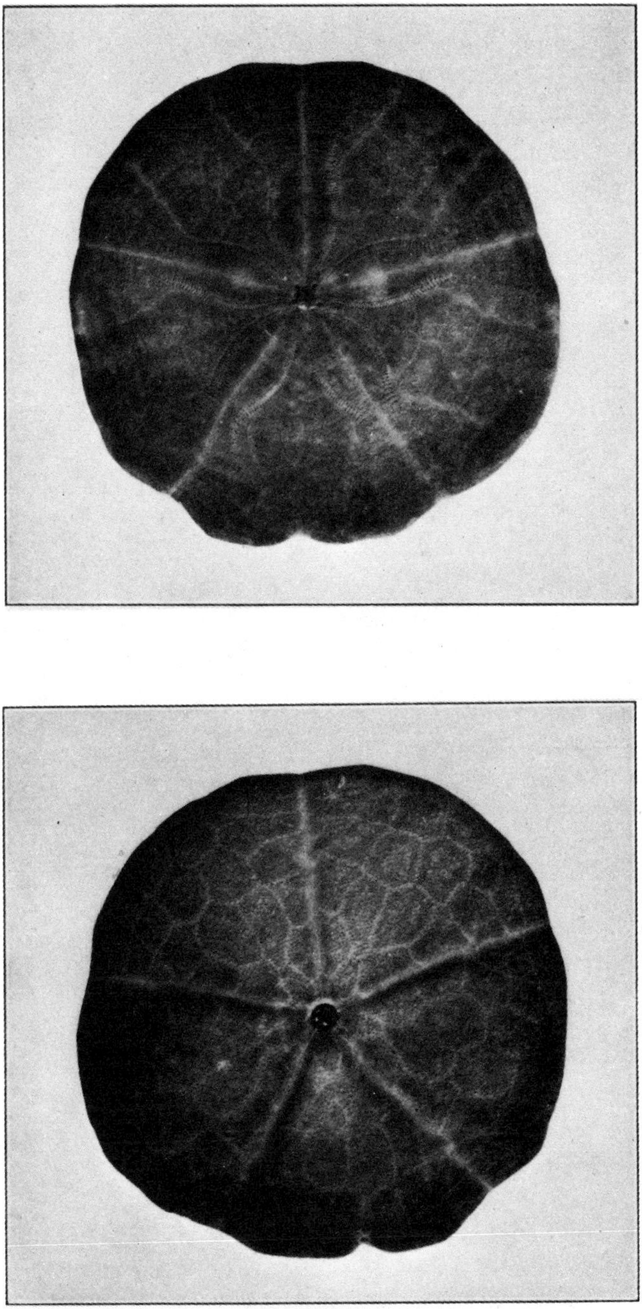

PLATE XXIII. Skeleton of Disk-urchin, *Echinarachnius parma.* (Natural size.)

here surround a single central plate, however, instead of the intestinal opening.

The ambulacral areas are plainly marked on the aboral surface of both species by the rows of respiratory tube-feet which are arranged symmetrically about the central plate. The figure thus formed has the general shape of a five-petaled flower (Plates XXVII and XXVIII), and the two groups of each ambulacral area are together called a "petal." The petals are even more clearly indicated by the rows of double pores on the surface of the disk after the removal of the spines (Fig. 14).

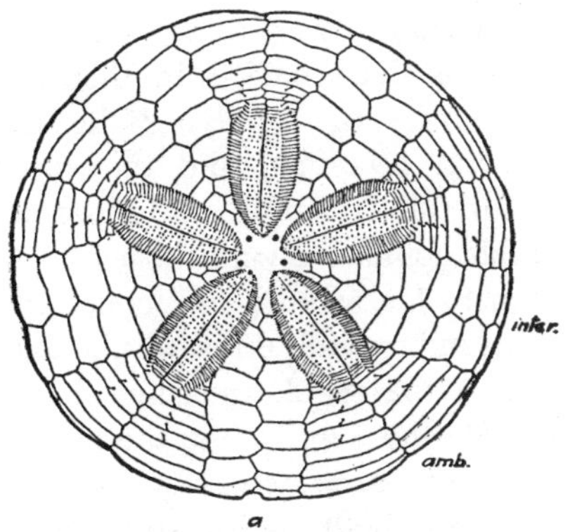

FIG. 14. *Echinarachnius parma.* Outline of plates on aboral surface; *amb,* ambulacral areas; *inter,* interambulacral areas; *a,* position of intestinal opening. The four genital pores are shown near the center of the disk.

In contrast with the large spines of our regular sea-urchins, the disk-urchins are clothed everywhere with minute, slender, brownish or purplish spines, those on the upper surface being covered with cilia, which cause currents of water to pass continually over the respiratory tube-feet.

The tube-feet in the clypeastroids are specialized either for respiration or locomotion. Those serving the latter function are scattered all over the upper surface in the interambulacral areas as well as in the ordinary position in the ambulacral areas. On the ventral surface they occur along definite grooves in the ambulacral areas, with branches extending into the interambulacral areas.

The respiratory tube-feet which mark the outlines of the petals, as stated above, are much larger than the others, and are bilobed at the base, with two fine tubes which pass through the plates of the test in the double pores of the petal. Both kinds of tube-feet bear suckers.

Correlated with this highly modified arrangement of the tube-feet is a corresponding specialization of the water-vascular system, for each foot is connected with a branch of these vessels.

MacBride* offers the following explanation of this complex system: "The condition of the water-vascular system is to be explained entirely by the peculiar environment of the animal. The demand for specialized respiratory organs is brought about by the habit of living half buried in the sand. Under these circumstances the strain of supplying the needful oxygen is thrown on the dorsal tube-feet, and they become modified in order to fit them for this function. The locomotor tube-feet are very small and feeble compared with those of *Echinus esculentus;* but this is comprehensible when it is recollected how little resistance the yielding sand would offer to the pull of a powerful tube-foot like that of the regular urchins, for in order to move the creature through the sand a multitude of feeble pulls distributed all over its surface is necessary, and the locomotor tube-feet are exactly fitted, both as to size and number, for this object." The power of movement is not wholly dependent upon the action of the tube-feet, however, for the vast number of minute spines which cover the body are freely movable and undoubtedly aid largely in the process of locomotion.

The structural weakness due to the flattened shape of the body in the disk-like urchins is compensated by strong vertical plates of calcareous material which unite the oral and aboral parts of the test so firmly, especially towards the edges, that it is difficult to break open the body except in its central portion.

Internal Anatomy.— The alimentary canal shows considerable modification in the disk-urchins, owing to the nature of the food and the position of the anal opening. Dissections showing the digestive tracts of the sand-dollar and the key-hole urchin are illustrated on Plate XXIV.

---

* Cambridge Natural History, Echinodermata, page 547.

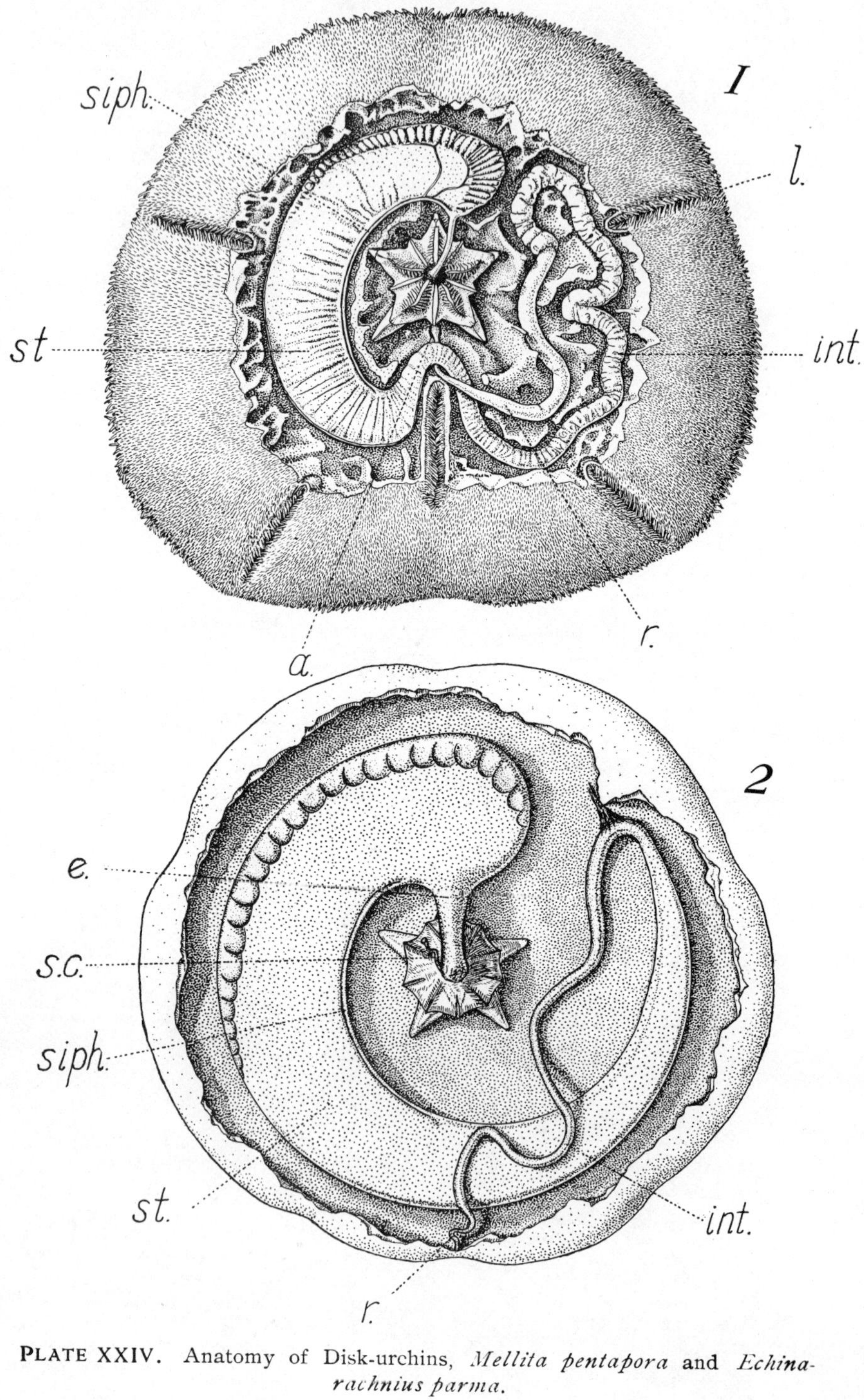

Plate XXIV. Anatomy of Disk-urchins, *Mellita pentapora* and *Echinarachnius parma*.

The reproductive organs are likewise modified, for the presence of the anus in the posterior interradius prevents the development of the gonad which in the regular urchins occupies this position. There are thus only two pairs of gonads developed in the adult with four symmetrical openings on the plates at the apical pole of the test, as shown in Fig. 14.

In the key-hole urchin the mouth lies near the center of the oral surface, as in the sea-urchin and sand-dollar, but the position of the anus is different from that which is found in any other echinoderm of the region, being likewise situated on the oral surface, and only a short distance posterior to the mouth (Plate XXIV, fig. 1).

The mouth is provided with five massive jaws, each ending in a sharply pointed tooth. The esophagus (Plate XXIV, fig. 1) leads towards the aboral surface, and then bends sharply anteriorly until it reaches the circle of supporting plates which surround the jaws. It then divides, as in *Arbacia,* into two branches, one of which forms the broad, flat stomach, while the other, which is closely attached to the inner margin of the stomach, is known as the siphon. The stomach (*st*) is just broad enough to occupy all the space between the inner circle of supporting plates and the outer system of plates which pass between the oral and aboral surfaces of the body.

The outer margin of the stomach is, as in *Arbacia,* provided with numerous sac-like lobes which greatly increase the secretory surface.

Explanation of Plate XXIV. Anatomy of Disk-urchins, *Mellita* and *Echinarachnius.*

Fig. 1. *Mellita pentapora.* Dissection of body to show alimentory canal. The slender esophagus leads from the centrally placed mouth (surrounded by five large jaws) to the broad stomach (*st*) and siphon (*siph*). The convoluted intestine (*int*), follows the definite course indicated, and passes gradually into the rectum (*r*), which opens through the anus (*a*) situated between the mouth and the median lunule. In some specimens the anus is placed at the inner end of the lunule. The five lunules (*l*) are shown.

Fig. 2. *Echinarachnius parma.* Dissection of the body to show alimentary canal; *e,* esophagus; *st,* stomach; *siph,* siphon; *int,* intestine; *r,* retum, opening on oral surface near margin of disk; *s. c,* madreporic canal. The five angular jaws are situated symmetrically about the mouth.

The stomach completes a half-circle of the body cavity in the left-handed direction, as seen from the aboral side, and is then joined by the narrow siphon mentioned above. At this point the intestine begins, and this organ bends sharply towards the mouth in order to pass between the jaws and the posterior lunule.

In the right half of the body the intestine bends sharply posteriorly beside the posterior lunule and runs in a series of broad loops nearly half-way around the body or nearly to the middle line anteriorly. It then bends sharply back on itself, occupying a position between the inner and middle series of supporting plates, and returns to the middle line between the mouth and the posterior lunule. On reaching a position near the middle line, the intestine changes into a narrow tube, the rectum (Plate XXIV, fig. 1, *r*), which crosses the first portion of the intestine on the aboral side, and then bends sharply to open to the exterior between the anterior border of the posterior lunule and the mouth.

The siphon referred to above as lying along the inner border of the stomach, is a narrow tube which makes a short passage between the two ends of that organ. By means of this siphon water can be taken in at the mouth and passed back directly to the intestine without interfering with the digestive juices of the stomach. The water thus introduced is said to serve a respiratory function in furnishing a supply of oxygen to the viscera.

In both the key-hole urchin and the sand-dollar, the posterior portion of the stomach and the whole of the intestine are commonly filled with a mass of fine sand and mud, abounding in diatom shells. This is sufficient indication that the food of these animals consists largely of such microscopic particles of organic matter as occur in the mud or adhering to the grains of sand. These particles doubtless include many kinds of living and dead organisms among which diatoms and other minute algæ are abundant.

If a fresh or well preserved sand-dollar be opened by carefully breaking away the plates of the aboral surface the internal organs are easily studied. The four brownish purple ovaries or spermaries are attached to the aboral plates. After these are removed the alimentary canal lies freely exposed.

The digestive system of the sand-dollar (Plate XXIV, fig. 2) is essentially similar to that described for the key-hole urchin, with the exception that the rectum opens to the exterior near the edge of the disk, and is on that account longer than in the latter species.

The short esophagus leads from the large jaws surrounding the mouth to the broad flat stomach. In life the stomach is green in color, with a smooth inner border (Plate XXIV, fig. 2), while its outer border is provided with many lobes of brownish color. A narrow siphon extends along the inner border, as in the key-hole urchin. The intestine nearly completes the wide circle, of which the stomach forms a part, and then bends sharply on itself to pass posteriorly, with many short bends, to the opening near the edge of the disk. The opening is sometimes almost exactly on the margin, but is more often placed near the edge of the oral surface. In young individuals it is on the aboral surface.

HABITS

The four species of Echinoids which are found in Connecticut waters exhibit much diversity of habit, associated with differentiations of structure. The two species belonging to the order Regularia — the purple sea-urchin and green sea-urchin — are provided with large spines for protection and locomotion, and live free upon the surface of rocks, sand, shells, and other objects found upon the bottom of the sea. Between tides they often inhabit the tide pools in the rocks, hiding away in crevices and beneath seaweeds, often clinging to the frond of the seaweed itself and sometimes drawing it over the body as a protective covering. The two species of clypeastroids — the sand-dollar and the key-hole urchin,— on the other hand, are covered everywhere with extremely minute spines, and they live partly buried beneath the surface of the sand.

Locomotion.— The regular sea-urchins move about from place to place both by the movements of the spines and by the action of the very long tube-feet. Both sets of organs commonly assist in the process. Certain species make the greatest use of the spines in locomotion, while the tube-feet of other forms are the most efficient. In exceptional cases the former are said to be able to move rapidly without making any use of the tube-feet, and the latter to walk with the tube-feet exclusively without touching the spines to the object on which they are moving. The sand-dollar and key-hole urchin move through the sand by the actions both of the minute spines and of vast numbers of sucking tube-feet on both upper and lower surfaces of the disk-like body. These hundreds of tube-feet attach themselves to the particles of sand, and by their combined action draw the body forward, the minute spines aiding in the process.

The regular urchins can easily right themselves from any position in the water by the combined action of tube-feet and spines. The disk-urchins, on the other hand, can only right themselves after they have buried their bodies in the sand.

Locomotion in any direction is accomplished with equal ease by the regular urchins, while the disk urchins ordinarily move with the portion of the disk which bears the intestinal opening backward.

The behavior of various species of echinoids under a great variety of conditions has been studied extensively by numerous investigators. In the popular book by Romanes,* and in a recent book by Von Uexküll,† may be found most interesting accounts of the performances of these fascinating creatures.

The bodies of all species must be kept clean, and for this purpose special provision is made. In both the green and the purple urchin the intestinal opening is in the center of the upper surface. Consequently when the refuse matter is discharged it tends to fall directly upon the surface of the body. In the green urchin this region is thickly studded with minute movable spines and pedicellariæ which can be moved in such a way as to pass any foreign particles from one to the next nearer the border of the test, so that such particles are quickly cast away from the body. The tube-feet also aid in the cleaning process.

Elizabeth and Alexander Agassiz‡ describe this action of the pedicellariæ in the following words:—"If we watch the sea-urchin after he has been feeding, we shall learn at least one of the offices which this singular organ performs in the general economy of the animal. That part of his food which he ejects

* Jellyfish, Starfish, and Sea-Urchins.

† Umwelt und Innenwelt der Tiere.

‡ Seaside Studies in Natural History, page 105.

passes out at an opening on the summit of the body, in the small area where all the zones converge. The rejected particle is received on one of these little forks, which closes upon it like a forceps, and it is passed on from one to the other, down the side of the body, till it is dropped off into the water. Nothing is more curious and entertaining than to watch the neatness and accuracy with which this process is performed. One may see the rejected bits of food passing rapidly along the lines upon which these pedicellariæ occur in greatest number, as if they were so many little roads for conveying away of the refuse matters; nor do the forks cease from their labor till the surface of the animal is completely clean, and free from any foreign substance. Were it not for this apparatus the food thus rejected would be entangled among the tentacles and spines, and be stranded there till the motion of the water washed it away."

In the purple urchin the aboral area is devoid of spines, and so smooth that refuse matter or foreign particles which fall upon it will be easily washed away by currents of water or by the movements of the animal.

In the disk-urchins the intestinal opening is placed at the edge of the disk or on the lower surface, so that the fecal matter does not fall upon the body. Furthermore the spines are covered with cilia, and these create a current of clean water which constantly flows past the animal and washes away any very minute foreign particles. The movements of the minute spines and tube-feet brush away any particles which are not thus removed.

Food.— The regular urchins feed largely upon seaweeds, but also devour any dead animal matter obtainable and supplement their diet by the diatoms and such other minute organisms as are found upon the bottom on which the animals live. Much sand and mud is swallowed for the organic matter in it. The disk-urchins swallow great quantities of sand, with such diatoms and other minute organisms as it contains.

Natural enemies.— Many kinds of fishes feed upon all our species of echinoids, in spite of their covering of spines, and the free-swimming embryos are subject to the attacks of still other species of fishes.

REPRODUCTION

In the class Echinoidea there are no species which are capable of asexual reproduction, such as occurs in some of the Asteroidea, the production of eggs which require fertilization being the only means of reproduction in this class.

The echinoids are of separate sexes, although it is impossible to distinguish the males from the females externally. Each female discharges vast numbers of eggs into the water during the breeding season of each year of her life. In one of the foreign species a large female may produce twenty million eggs yearly. Owing to the gregarious habits of the animals, the eggs stand a reasonable chance of being fertilized by the sperm cells which are discharged in enormous numbers by the male urchins at the same season.

It often happens, as is also the case with the starfish, that certain animals fail to mature and discharge their sexual products at the normal season, but eggs discharged at other periods of the year must almost always fail of fertilization or development.

The eggs are so numerous and easily obtained that they have become classical objects for the study of the processes of fertilization and early development. The suggestions on page 50 regarding the study of starfish eggs will be found equally appropriate for those of the urchins.

DEVELOPMENT

As illustrated by Fig. 15, the immature eggs in the ovary of the urchin are each provided with a large germinal vesicle. After the two polar bodies are formed, the egg is discharged into the water; it then settles slowly to the bottom and is ready for fertilization. A single sperm cell, which may be swimming about in the water in the vicinity, may enter the egg. After such union the fertilized egg divides regularly into 2, 4, 8, and 16 cells, called blastomeres, as illustrated in Fig. 15.

These divisions take place very rapidly, the first cleavage occurring in about an hour and a quarter after fertilization, and succeeding cleavages at intervals of an hour or less.

A hollow sphere of cells known as the blastula is then formed. In about twelve hours after fertilization the blastula becomes

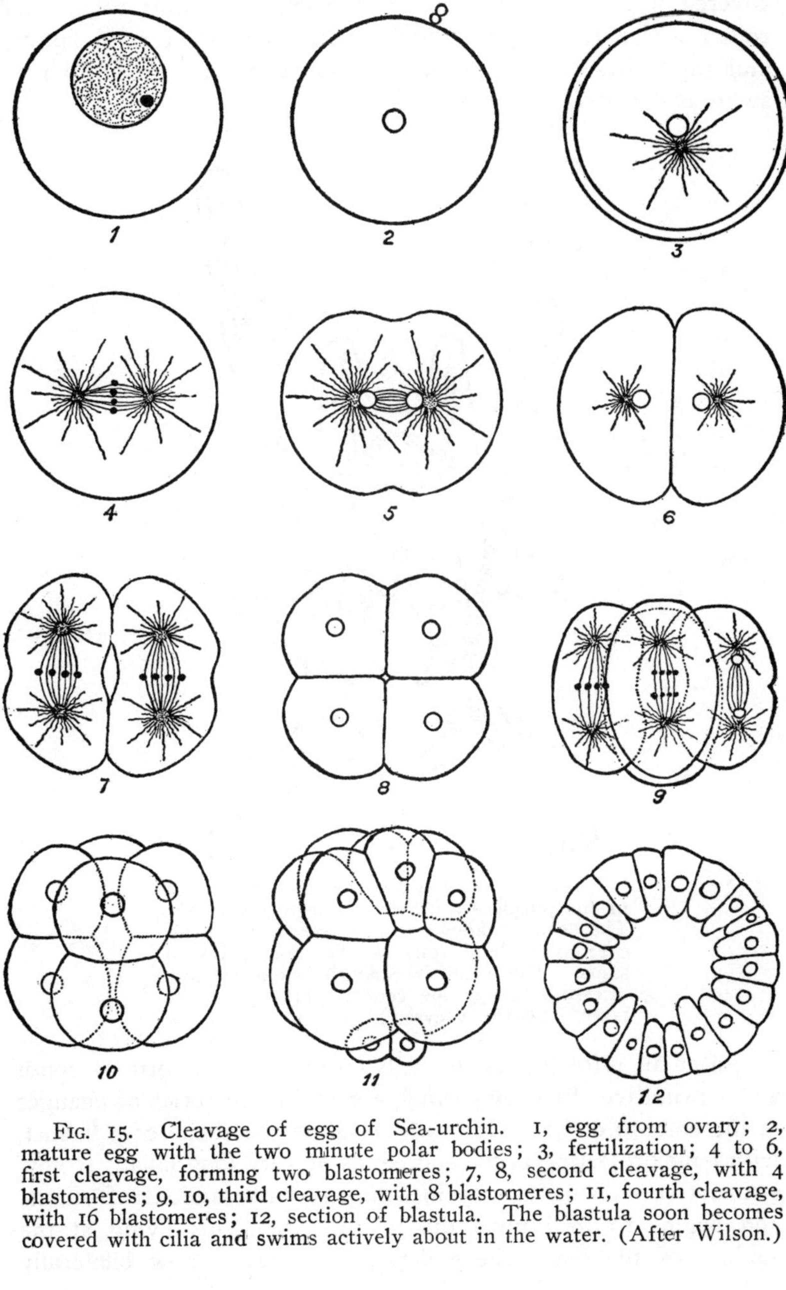

FIG. 15. Cleavage of egg of Sea-urchin. 1, egg from ovary; 2, mature egg with the two minute polar bodies; 3, fertilization; 4 to 6, first cleavage, forming two blastomeres; 7, 8, second cleavage, with 4 blastomeres; 9, 10, third cleavage, with 8 blastomeres; 11, fourth cleavage, with 16 blastomeres; 12, section of blastula. The blastula soon becomes covered with cilia and swims actively about in the water. (After Wilson.)

covered with vibrating cilia, and the embryo swims round and round within the egg membrane. This membrane soon bursts, and the free-swimming blastula rises from the sea bottom to swim at the surface of the water.

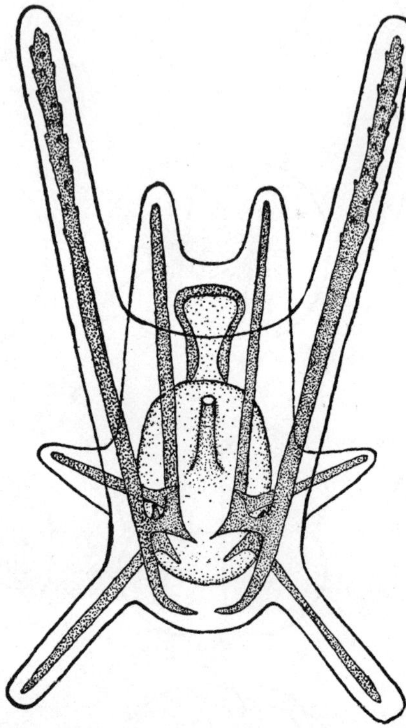

FIG. 16. Free-swimming larva (pluteus) of the purple sea-urchin, *Arbacia punctulata*. The deeply shaded portions represent the calcareous skeletal rods in the four pairs of arms. The primitive mouth, stomach, and intestine occupy the central portions of the body. (After Weysse.)

After an infolding of the cells on one side to form a mouth and a primitive alimentary canal, a complicated series of changes in the embryo occurs. It finally assumes the shape of a helmet, with long projecting arms, and bands of very long cilia. It now swims freely about, usually in company with millions of others, and begins to feed upon the minute organisms living at the surface of the sea. The embryo, or larva, is now bilaterally symmetrical, with both mouth and anus, and is known as a *pluteus* (Fig. 16). It bears absolutely no resemblance to the adult urchin. The free-swimming pluteus lives and feeds at the surface for a period of several weeks, the time depending somewhat on the temperature and other conditions, and varying in different species. In this time it may have been carried by currents and tides to a considerable distance from its place of origin.

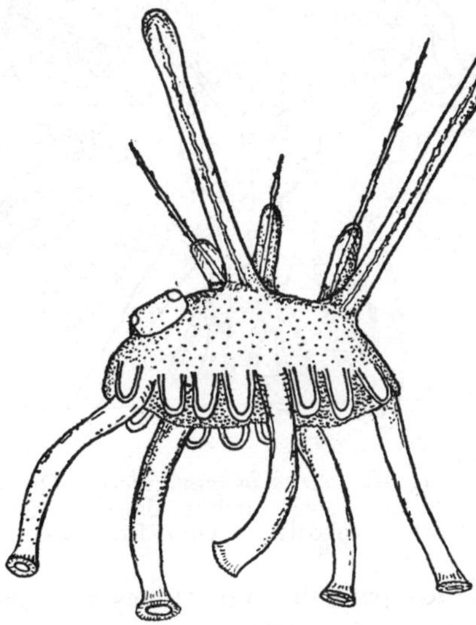

FIG. 17. *Arbacia punctulata.* Side view of young urchin immediately after the metamorphosis. The spines remaining from the pluteus are in process of absorption. (After Brooks.)

When ready to assume the adult form, the pluteus settles to the bottom, absorbs its larval organs, and by a complicated metamorphosis assumes the condition of a young urchin (Figs. 17 and 18). The young urchin differs from the adult in having the whole upper surface covered by the plates which will eventually form only the periproct of the adult, and by the relatively enormous tube-feet. The animal is now radially symmetrical, and creeps about on the sea bottom after the manner of the adult urchin. It feeds upon diatoms and other minute organisms until, by its further growth, and the development of plates and spines, it is able to assume the habits of the adult.

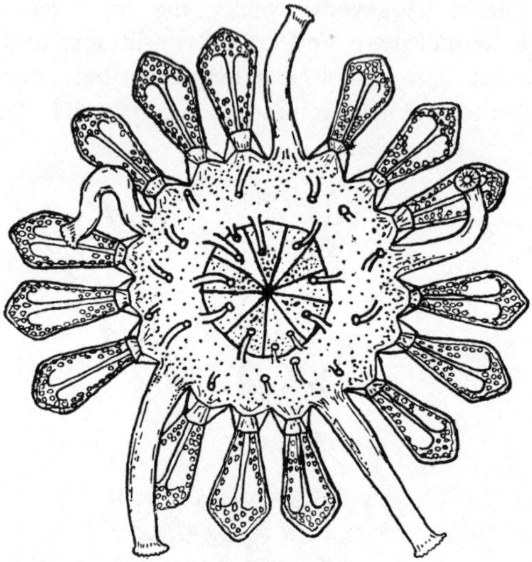

FIG. 18. *Arbacia punctulata.* Oral surface of young urchin shortly after the metamorphosis. (After Brooks.)

The **clypeastroids** pass through stages very similar to those of the regular sea-urchins. The sand-dollar, for example, develops from a long-armed pluteus, which settles to the sand at the sea bottom and gives rise to the young sand-dollar. This has at first a few relatively long spines and large tube-feet. There is now a nearly complete radial symmetry, the intestinal opening at this stage being near the center of the upper surface, somewhat as in the regular urchins. With the further growth of the animal, its bilateral symmetry becomes more marked, and the anus is gradually shifted to its final position near or beneath the edge of the disk.

Grave* has been successful in rearing large numbers of the key-hole urchin from the fertilized eggs, and following their transformations to the adult form. The egg of this species

* C. Grave. *Biol. Bulletin,* vol. v, 1903.

likewise develops into a long-armed pluteus, and this by a complicated metamorphosis gives rise to the bizarre creature shown in Fig. 19, which actually represents the young urchin, although it

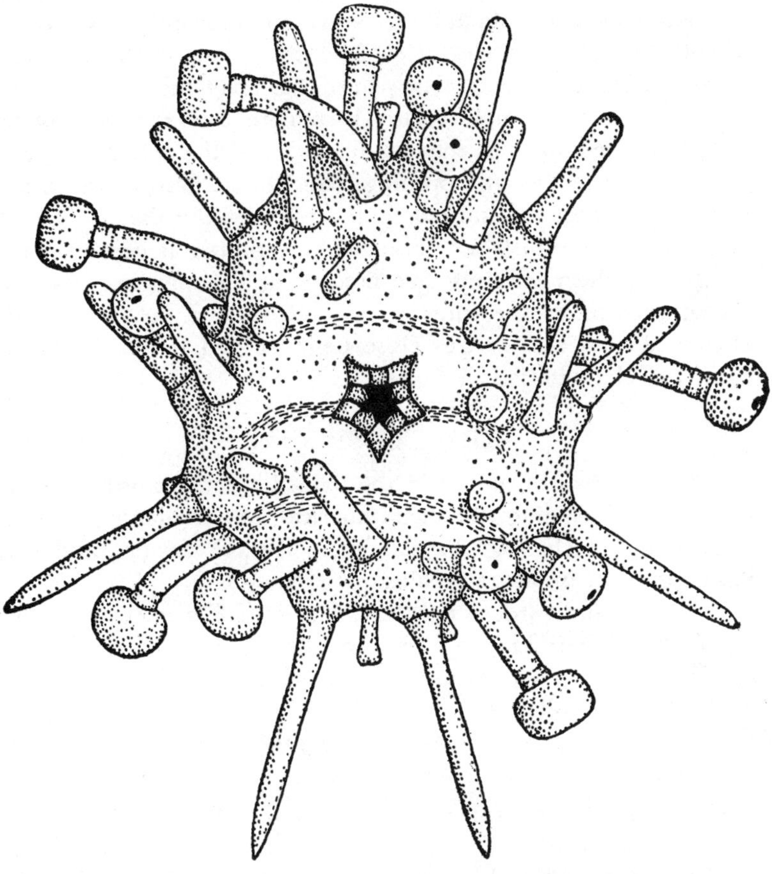

FIG. 19. *Mellita pentapora.* Young urchin shortly after the metamorphosis from the free-swimming pluteus. The pentagonal area in the center represents the mouth, in the angles of which are the five teeth. The spines and tube-feet are at this stage of enormous size relative to the size of the body. (After Grave.)

differs widely in appearance from the adult. The oval body is provided with a small number of spines and tube-feet of relatively enormous size. By means of the tube-feet the creature clings to the objects on which it rests and crawls slowly about. In the center of the body is the pentagonal mouth with the five calcareous teeth, and extending across the body are three bands of cilia which are thought to serve in sweeping towards the mouth the small particles of organic matter on which the animal feeds. With the increase in size of the body the ciliated bands disappear, and the now relatively minute tube-feet convey the food to the mouth. As the disk-like body of the animal increases further in size, the anal opening, which is at first on the aboral surface, is gradually shifted to the edge of the disk. The enlargement of the disk now becomes interrupted at certain points on the periphery, causing five notches in the margin. By continued growth of the intervening portions the notches become elongated, and their edges close together and unite distally, to form the characteristic lunules.

KEY TO SPECIES

The echinoids of the world are divided into three orders: (1) **Regularia,** including the hemispherical sea-urchins; (2) **Clypeastroidea,** including the flat and disk-like forms; and (3) **Spatangoidea,** or heart-urchins.

The two first orders only are found in Connecticut waters, and of these there are four representatives; two, the purple sea-urchin and the green sea-urchin, belonging to the Regularia, and two, the sand-dollar and the key-hole urchin, to the Clypeastroidea. They may be easily distinguished by the following characters:—

1. Body nearly hemispherical, spines rather large, anal opening at aboral pole 2
   Body flat and disk-shaped, spines small, anal opening eccentric 3
2. Spines long, absent near aboral pole... **Arbacia punctulata**
   Spines short, covering aboral pole.......................
   Strongylocentrotus dröbachiensis
3. Test without perforations.......... **Echinarachnius parma**
   Test with five narrow perforations (lunules).............
   Mellita pentapora

PLATE XXV. Purple Sea-urchin, *Arbacia punctulata.* Tests with spines removed. (Natural size.) After Clark.

Arbacia punctulata (Lamarck)

Purple Sea-urchin

Plates XX, XXII, XXV.

This is the only species of the regular sea-urchins common in the greater part of Long Island Sound. It is widely distributed, but occurs in abundance only in restricted localities. It prefers shelly or rocky bottoms, and is often dredged from the oyster beds. It sometimes occurs between tides in the tide pools in rocky places, hiding in crevices or beneath seaweeds.

The species occurs in shallow water along the coast from Cape Cod to the Gulf of Mexico. There is great variation in the abundance of individuals of this species in the same locality in different years. At the Biological Laboratories at Woods Hole, Mass., large numbers of these urchins are desired for embryological and experimental studies each year, and, while in some years the species can be found in unlimited numbers, in other years it is extremely difficult to secure a supply. On the oyster beds off New Haven Harbor, there seems to be a similar variation in numbers.

The purple sea-urchin is commonly from one to two inches in diameter, excluding the spines. The spines are heavy and solid, and measure from one-half to one inch in length. Each is fluted longitudinally, and, while some of the upper ones are acutely pointed, others are blunt. Towards the oral surface they become shorter and flattened at the tip. They are usually of a purplish color. The test is likewise reddish, purple, or purplish brown, while the tube-feet are red or brown.

EXPLANATION OF PLATE XXV. Purple Sea-urchin, *Arbacia punctulata*. (Natural size.)

Fig. 1. Aboral surface of test with spines removed, showing arrangement of plates. The five double rows of dotted plates represent the ambulacral areas, while the intervening rows of plates comprise the interambulacral areas. The four plates at the center form the periproct. Each of the interambulacral areas ends in a plate with a small perforation, the genital pore. One of these plates is somewhat larger than the others and serves as the madreporic plate.

Fig. 2. Side view of test.

Fig. 3. Oral surface of test, showing the large opening covered in life by the peristome.

(From photographs loaned by the U. S. Fish Commission, with permission of Dr. H. L. Clark.)

The aboral pole is free from spines for a considerable distance; not only are the plates of the periproct and the genital plates naked, but spines are likewise absent from the middle of the interambulacral areas for a distance nearly half-way to the equator of the test. Only the large primary spines are present. They are arranged in a double series in each ambulacral area, and in four to eight or more series in each interambulacral area. There are five pairs of prominent buccal plates.

In Connecticut waters the sexual products are matured in early summer. The eggs contain particles of the purplish pigment which permeates nearly all the tissues of the body. They are admirable objects for the study of the processes of fertilization, cleavage, and early embryonic development.

For such a study the suggestions given for the examination of the starfish egg will apply equally well. The free-swimming embryo, known as the pluteus, may be kept alive for a number of days, but nearly all attempts to rear artificially to the adult form have failed.

Strongylocentrotus dröbachiensis (O. F. Müller)

Green Sea-urchin

Plates XXI, XXVI.

This is a typical circumpolar species, occurring in great abundance north of Cape Cod, and is found but locally in Connecticut waters. It extends southward in the deep water off our coast as far as New Jersey, but south of Cape Cod it is of much smaller size than northward.

In Long Island Sound small specimens have been dredged off New Haven and Stratford; but these individuals or their ancestors have presumably been carried in their early stages by the

---

Explanation of Plate XXVI. Green Sea-urchin, *Strongylocentrotus dröbachiensis.* (Natural size.)

The upper figure shows the aboral surface, thickly covered with the primary and secondary spines.

In the center of the aboral surface (lower figure) is seen the peristome with the five sharp white teeth which close the mouth. Between the spines are shown the five double rows of slender tube-feet with enlarged terminal sucking disks.

(Photographs loaned by the U. S. Fish Commission, with permission of Dr. H. L. Clark.)

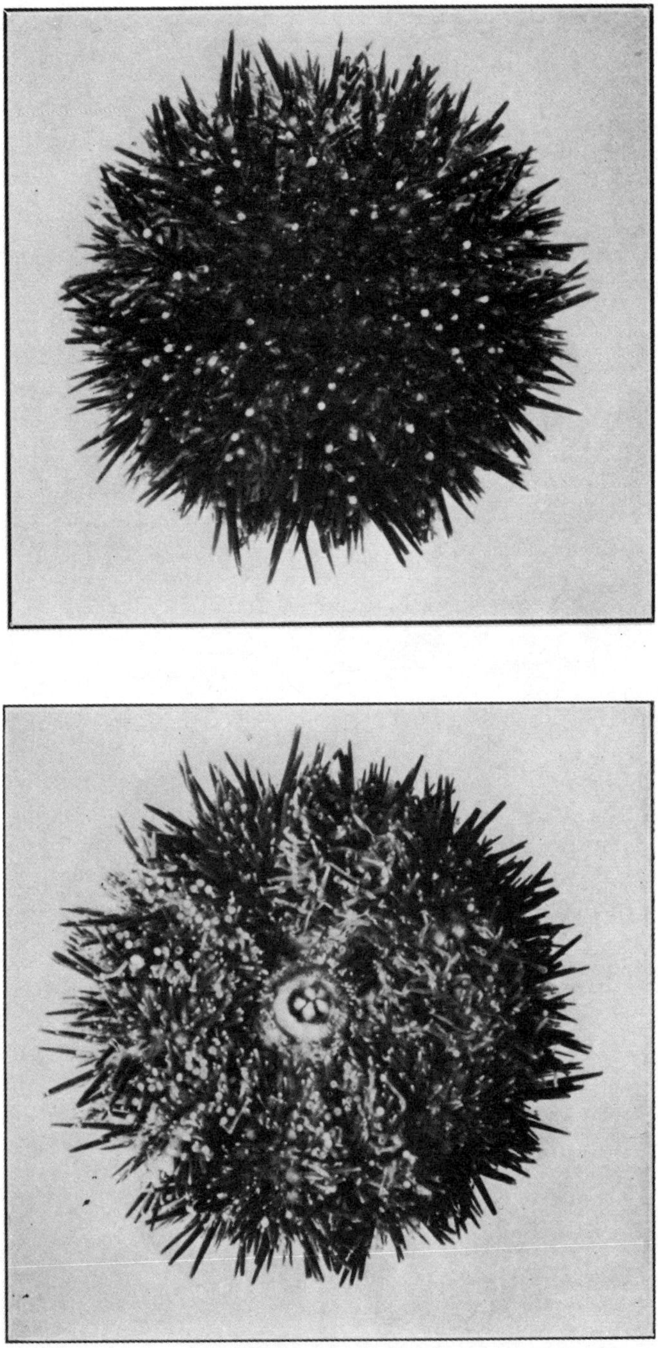

PLATE XXVI. Green Sea-urchin, *Strongylocentrotus drö-bachicnsis*. (Natural size.)

westward currents from the colder and deeper waters nearer the eastern portion of the Sound. Specimens have been taken near the Thimble Islands and near Faulkner's Island. Off New London and eastward to Watch Hill, Rhode Island, the individuals become increasingly abundant and of larger size. The species also occurs on the northern coasts of Europe, the north coast of Siberia, and on both sides of the northern Pacific Ocean and Bering Sea, southward to Kamchatka on the west and Puget Sound on the east.

The species prefers rocky bottoms, the food consisting largely of algæ growing upon the rocks. Mud containing diatoms and other minute organisms is swallowed in abundance, and any dead animal is greedily devoured. Even the bones of a dead fish will be cut away with the sharp teeth and completely devoured.

Farther to the north the species lives in great abundance between tides. In the southern part of its range it may reach a depth of more than 600 fathoms.

Individuals of this species grow to a larger size than do those of the purple urchin, the diameter of the test measuring from two to three inches in large specimens.* The test is rather flattened, being about half as high as broad. The spines are exceedingly numerous, but are small and short, the largest being only about a half-inch in length.

The peristome measures from three-fourths to one inch in diameter, and is proportionally larger in young than in mature individuals. The anal plates are numerous and very small in full-grown specimens.

The color in life is greenish, varying in shade in different individuals and often on different parts of the body. Many specimens have the zones bearing the tube-feet much lighter in shade than the interambulacral areas. In many there is a yellow, red, or purple tinge, due to variations in the coloration of the spines; and in some these shades are intensified, the spines being tipped with deep red, purple, or violet. The miliary spines, pedicellariæ, and tube-feet are whitish or pale violet.

The species differs from *Arbacia punctulata* in that all the plates of the aboral pole bear minute protective spines. The larger spines of the test are longitudinally fluted and pointed at the tip. They are arranged in a double series on each ambulacral

* Several very large specimens in the Yale University Museum exceed four inches in diameter.

and interambulacral area. Scattered thickly among the primary spines are much smaller, secondary, spines of similar shape, while a third variety, the miliary spines, are still smaller and very slender. There are five pairs of buccal plates, which bear miliary spines. The comparative sizes of the spines and their arrangement are well shown by the tubercles on the test (Plate XXI).

Several kinds of large fishes, including the wolf-fish, swallow the adult animals whole, quite disregarding the sharp spines which cover the body. The young have still more numerous enemies among the fishes, and the free-swimming plutei are swallowed in great numbers by surface-feeding fishes.

Echinarachnius parma (Lamarck)

Sand-dollar

Plate XXIII; Plate XXIV, fig. 2; Plate XXVII.

The sand-dollar is a most interesting little creature because of its peculiar modifications to meet the requirements of the environment under which it lives. It is sometimes very abundant on sandy bottoms off shore, where it lives partly buried beneath the surface of the loose sand. The body is covered with a multitude of tiny suckers, by means of which the creature creeps slowly along, devouring as it goes particles of sand with the numerous microscopic organisms, particularly diatoms and other algæ, which are found among them.

The much flattened, disk-like body is well adapted for withstanding any tendency to be rolled over or thrown upside down by the shifting of the sand under the influence of the waves. It is very probable, moreover, that the animal can penetrate deeper into the sand when the water becomes very rough, to return to the surface again after the disturbance is over.

The species has a remarkable range of distribution. It is found along the Atlantic coast from Labrador to New Jersey, and

---

Explanation of Plate XXVII. *Echinarachnius parma.* Aboral and oral surfaces. (Natural size.)

The aboral surface, upper figure, shows the minute spines and the darker ambulacral areas, "petals". The oral surface, lower figure, shows the ambulacral areas as narrow furrows radiating from the centrally placed mouth and branching on each side toward the margin. These areas are nearly devoid of spines. A cluster of less slender spines surrounds the mouth.

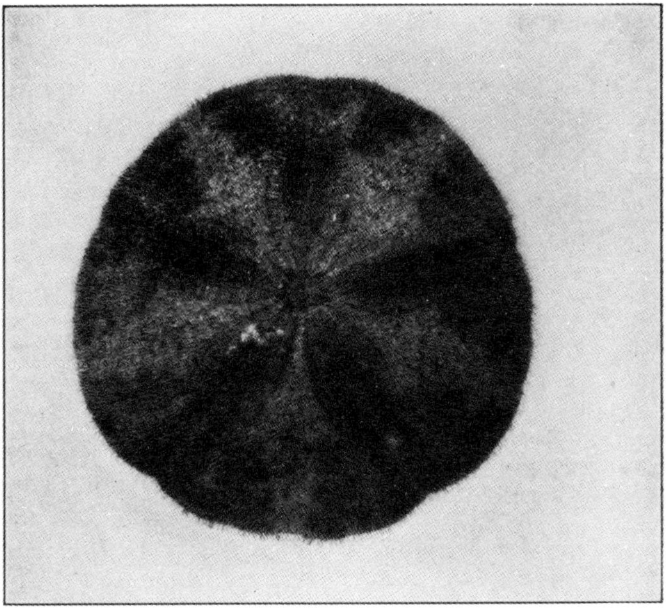

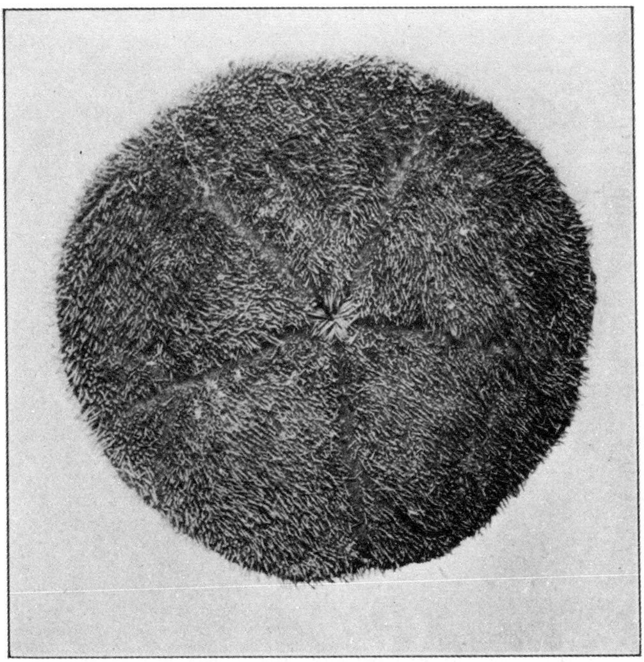

PLATE XXVII. Sand-dollar, *Echinarachnius parma*. Aboral and oral surfaces. (Natural size.)

on both sides of the Pacific Ocean, extending from the Aleutian Islands along the American shore as far south as British Columbia and on the Asiatic shore southward to Japan. In the more northern parts of its wide range it is found near low-water mark, but farther south it occurs only in deeper water. It has been collected from a depth of more than eight hundred fathoms.

The color of the body in life is usually purplish brown or reddish brown, some specimens inclining to purple while others are distinctly red. When young, a paler red color predominates. Many individuals have the ambulacral areas darker red than the remainder of the body. In alcohol, the color changes to a dark green or brown, and the alcohol dissolves so much of the pigment as to become deeply colored. Dried specimens are likewise dark green or brown. Verrill states that the fishermen sometimes made an indelible marking ink by grinding up the skins and spines of these animals and mixing with water.

The body is nearly circular in outline, but is indented at one point, which indicates the position of the anus. The surface of the body is everywhere thickly covered with minute spines (Plate XXVII), so delicate and closely placed as to form a velvety covering.

On the aboral surface the ambulacral areas are plainly marked by the position of the respiratory tube-feet. These areas are symmetrically placed about the central plate, and constitute the so-called petals.

In the test after the removal of the spines (Fig. 14), the position of the petals is clearly shown as forming the aboral ends of the ambulacral areas (*amb*). The interambulacral areas likewise bear tube-feet, but they are here much smaller than those of the petals. At the tips of the ambulacral areas the position of the terminal tentacle is indicated, and between them the four openings of the genital glands.

On the oral surface the ambulacral areas are marked by furrows (Plate XXVII). The peristome is very small (Plate XXIII), although the five calcareous teeth are well developed and very sharp.

In many localities the species is so abundant as to form an important part of the food supply of certain fishes, particularly of the flounder, codfish, and tautog.

Mellita pentapora (Gmelin)

Key-hole Urchin

Plate XXIV, fig. 1; Plate XXVIII.

The name *Mellita testudinata* was given to this species by Klein in 1734; but, as this long preceded the Linnæan code, the specific name *pentapora* proposed by Gmelin in 1788 should be adopted. Many authorities, however, insist upon the use of *testudinata.**

This species has a flat test, and in this respect bears a superficial resemblance to the common sand-dollar, but is easily distinguished by the five narrow openings, termed lunules, which pass completely through the body as shown on Plate XXVIII. One of these lunules is situated in the posterior interradius, about half-way from center to circumference, while the other four are symmetrically placed, one in the center of each radius except the anterior.

The flattened test is densely covered with short but slender spines of a brownish color. The general color of the living animal is commonly pale brown, sometimes with a tinge of green.

Especially after the spines are removed, the petaloid ambulacral areas are conspicuous on the aboral surface. On the oral surface the narrow ambulacral furrows are provided with paired branches which extend peripherally from the mouth to the edge of the disk. Each of the main branches has lateral offshoots (Plate XXVIII).

* In a recent paper (*Annals and Mag. Nat. Hist.*, 8 ser., vol. vii, 1911) Clark recommends the adoption of the name *quinquiesperforata* Leske for this species.

Explanation of Plate XXVIII. Key-hole Urchin, *Mellita pentapora.* (Natural size.)

Aboral and oral surfaces of two dried specimens. The minute spines have been removed from the left-hand side of each. The aboral surface shows three of the five ambulacral areas, "petals". The mouth lies in the center of the oral surface, with the anus near the inner end of the median lunule. One of the posterior lunules in the lower figure is abnormal in that it remains open to the edge of the disk.

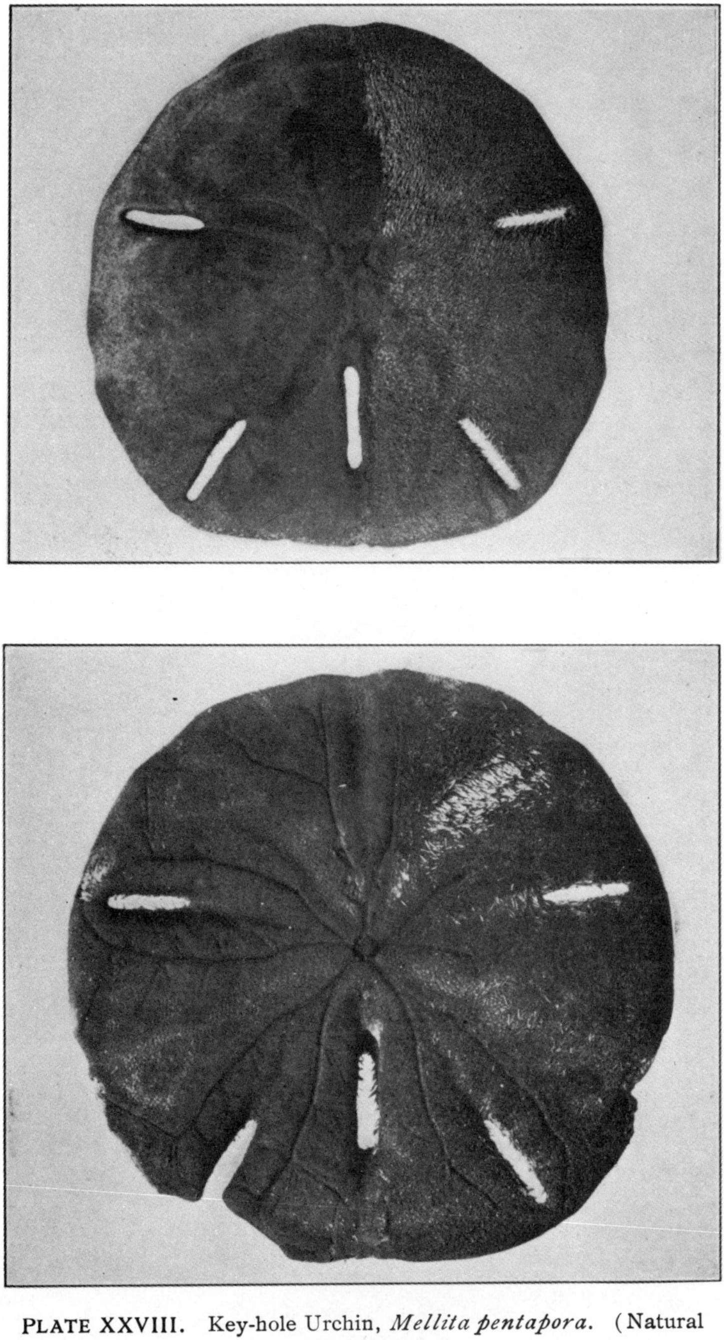

PLATE XXVIII. Key-hole Urchin, *Mellita pentapora.* (Natural size.)

The mouth is situated a little anterior to the center of the oral surface, while the posterior opening of the intestine lies between the central end of the interradial lunule and the mouth.

In some specimens one or more of the lunules, as shown on Plate XXVIII, may extend quite to the margin of the disk. Such an abnormality may be caused either by accidental injury or by the failure of the two edges of the lunule to unite together distally in the young urchin.

This species occurs locally from Nantucket to Brazil in shallow water, although it is not common north of Cape Hatteras. In Long Island Sound the species has been collected at Huntington, Long Island, a number of specimens from this locality belonging to the Yale University Museum.

A description of the digestive system of this species may be found on page 95, while Plate XXIV, fig. 1, illustrates the internal anatomy.

In a paper by Grave* the structure and development of the larvæ have been described and illustrated.

---

* *Johns Hopkins Univ. Circular*, No. 157, 1902.

CLASS IV. HOLOTHUROIDEA

The animals included in this class are sometimes called sea-cucumbers. They are, however, not so well known to people other than naturalists as are the other classes, and most of them can hardly be said to have any English names.

The six species of holothurians found along the Connecticut coast or in reasonable proximity thereto are all limited to certain restricted localities, and their presence in any particular spot cannot be predicted with any considerable degree of certainty. In general, a muddy bottom below low-water mark is favorable for the large, brown or blackish *Thyone briareus,* while the others prefer sandy localities, the two species of *Synapta* occurring frequently between tides, and the *Cucumaria* only below the ordinary tides. The two other species, *Thyone unisemita* and *Thyone scabra,* have not yet been recorded from Connecticut waters, but their presence in Narragansett Bay and near the eastern end of Long Island Sound makes it probable that they actually occur within the limits of the Sound, and that future investigations will reveal their presence in these waters.

STRUCTURE

These animals have but little superficial resemblance to the other classes of Echinoderms, although in their general anatomical structure they show their relationship. In contrast with the

---

EXPLANATION OF PLATE XXIX.

Fig. 1. *Synapta roseola.* Five times natural size. The two longitudinal lines represent two of the five bands of longitudinal muscles.
Fig. 2. Tentacle of same species, with seven digits. Much enlarged.
Fig. 3. *Synapta inhærens,* with the twelve tentacles surrounding the widely opened mouth. The body is strongly contracted and shows the constrictions due to action of the circular muscles. Along the axis of the body one of the bands of longitudinal muscles is shown. Five times natural size.
Fig. 4. Tentacle of same species, with nine digits. Much enlarged.
Fig. 5. *Thyone briareus.* Two-thirds natural size. The dorsal surface is toward the top of the page, the lower, rounded margin being in the mid-ventral line.

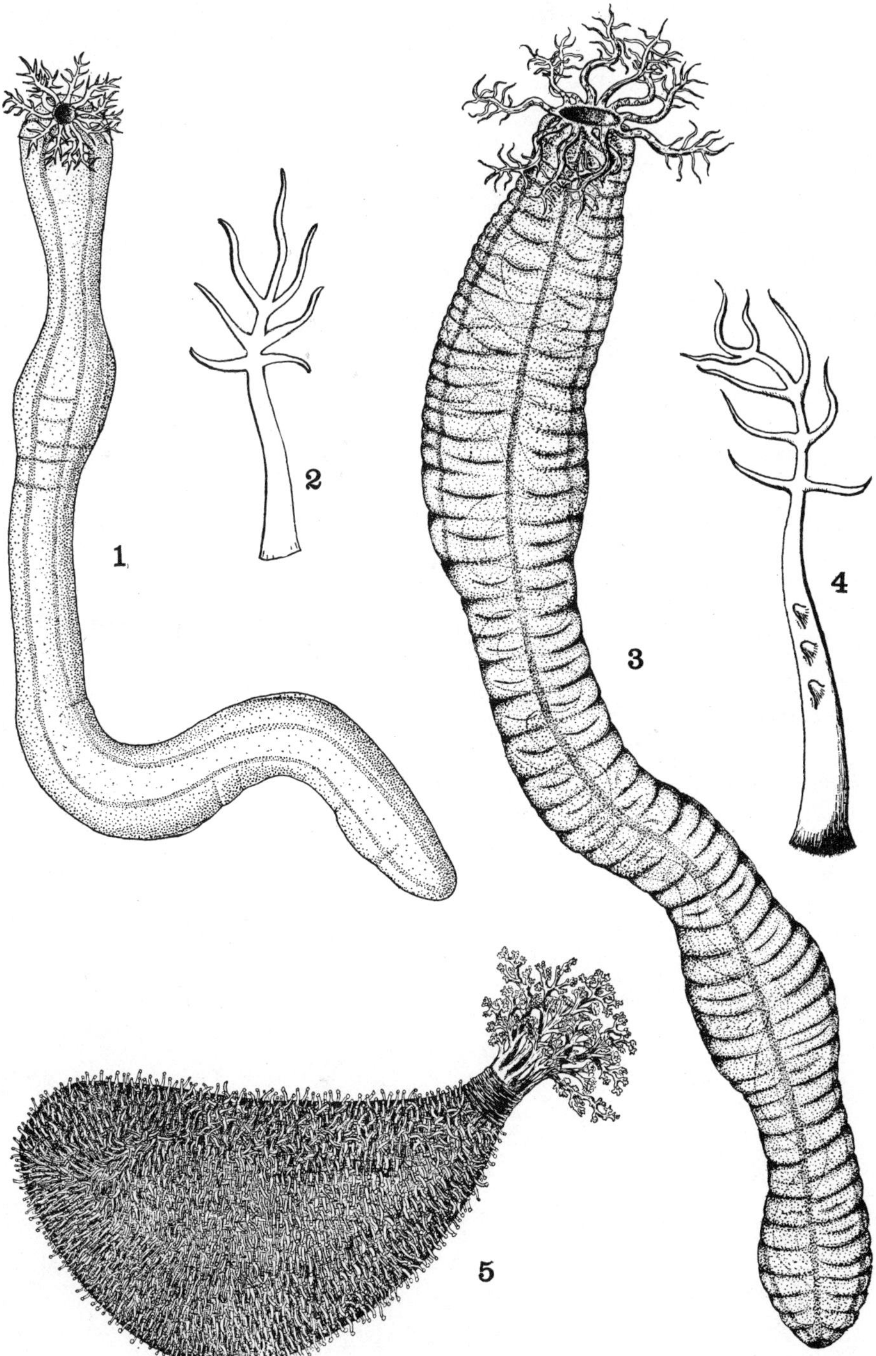

PLATE XXIX. Holothurians.

hard, spiny body coverings possessed by the starfish, brittle stars, and sea-urchins, the holothurians have soft, sac-like or worm-like bodies (Plates XXIX, XXXI, XXXII), in the skin of which only very small calcareous plates are found. The mouth, at one end of the elongated body, is surrounded by a circle of branched oral tentacles comparable with the oral tube-feet or tentacles of the other Echinoderms. The intestine ends in a sac-like cloaca at the end of the body opposite the mouth.

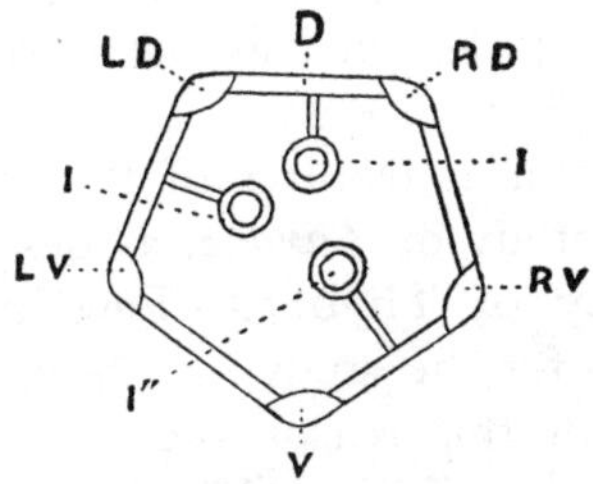

FIG. 20. Diagram of transverse section of the body of a holothurian. *D,* dorsal interradius; *I, I', I'',* first, second, and third sections of intestine attached to adjacent interradii; *LD, LV, RD, RV,* and *V,* left dorsal, left ventral, right dorsal, right ventral, and ventral radii, respectively.

The body is in most species adapted for creeping upon the sea bottom; and, as the animal habitually creeps with the same side uppermost, there has come about a fairly well-marked differentiation into a ventral and a dorsal surface.

Three of the five ambulacral areas are situated upon the ventral surface, and are usually provided with more highly developed tube-feet, or pedicels, with sucking disks. The pedicels of the two dorsal ambulacral areas are commonly modified as delicate finger-like processes used for respiration and sensation. In many forms, as *Thyone* (Plate XXXI), the pedicels are scattered over the whole surface of the body, while in such burrowing forms as *Synapta* there are no pedicels whatever (Plate XXIX, figs. 1, 3).

With the habit of creeping upon the side there has also come a bilateral symmetry of the body, so that it is customary to speak of the oral and aboral ends as anterior and posterior, and the terms right and left are applied as in other bilaterally symmetrical animals.

The body walls are very muscular, with enormously powerful longitudinal muscles running the length of the body along the five radii, which correspond to the ambulacral areas of the other groups (Fig. 20). Transverse muscles serve to diminish the circumference of the body, and thus, by exerting pressure on the fluids in the body, cause the animal to elongate. The interaction of the two sets of muscles enables the creature to move in a worm-like fashion.

The general anatomical features of the holothurians may be well illustrated by a study of *Thyone,* as described below.

Internal Anatomy of Thyone.— Few marine invertebrates are more satisfactory for the study of internal anatomy than the common *Thyone.* For this reason the more conspicuous of its organs will be briefly described in this place, although the details of the water-vascular system, blood system, and nervous system, the description of which would require more space than is here available, must be omitted.

So powerful are the contractions of the body walls when the animals are injured, that, unless some special precautions are taken in killing the specimens for dissection, the body may be ruptured, and the whole crown of tentacles and a portion of the viscera may be ejected. A convenient method of avoiding this trouble is to seize the living holothurian immediately behind the tentacles with strong forceps and plunge it into boiling water. If the experiment is successful, the tentacles will remain partially expanded, and the body walls intact.

For a study of the internal organs, the body should be opened with scissors along the whole length of the ventral surface.

On opening the body, the viscera are found to be situated in a large central cavity, the body cavity, or cœlom. In life the space between the viscera and the body walls is filled with the cœlomic fluid.

Alimentary canal.— Of the ten branched tentacles which surround the mouth, two are much smaller than the others, and will serve to indicate the ventral side of the body (Plate XXX). If a slit is made through the ventral body wall throughout the

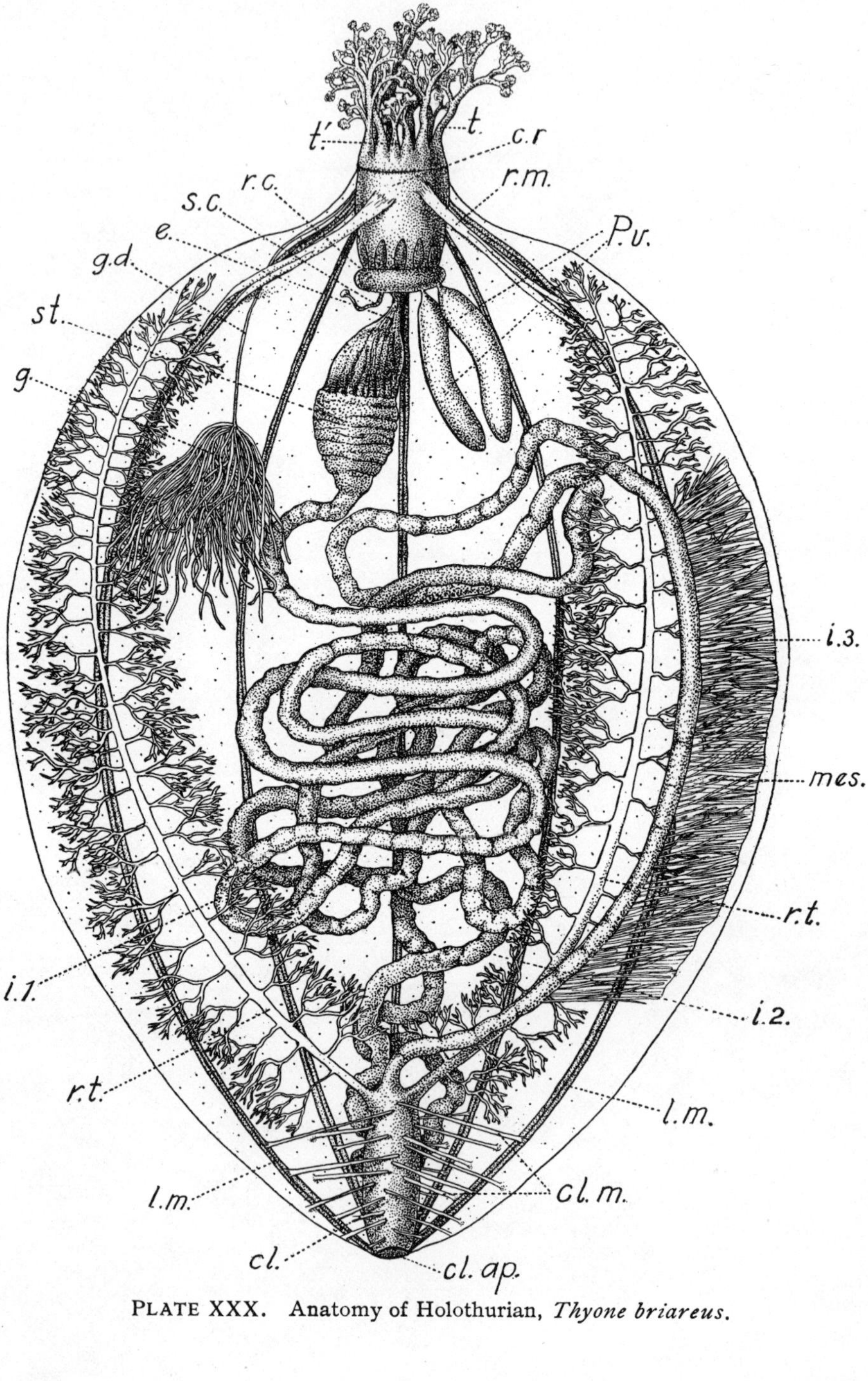

PLATE XXX. Anatomy of Holothurian, *Thyone briareus*.

length of the body, and the cut walls folded and pinned back, the arrangement of the internal organs is plainly visible.

In the spacious body cavity lie the much coiled intestine, the profusely branched respiratory trees, the Polian vesicles, the madreporic canal, and other organs, as shown on Plate XXX.

At the anterior end of the body the slender esophagus leads from the region of the calcareous ring at the base of the tentacles back to the sac-like stomach. The walls of the stomach are highly muscular and much thicker than the other parts of the alimentary canal. At its posterior end the stomach becomes much narrowed, and leads into the long and much coiled intestine. The total length of the intestine is several times that of the body. It is arranged in three broad loops, each of which is in itself much convoluted (Plate XXX). The first section leads with many folds posteriorly (near the mid-dorsal line) to the dorsal side of the cloaca. Here there is a sharp bend with several convolutions, and the second section leads forward (on the left dorsal side) to the anterior half of the body. After several further twists back and forth through the cœlomic space, the third section leads posteriorly (on the right ventral side) to

---

Explanation of Plate XXX. Anatomy of Holothurian, *Thyone briareus.* (Twice natural size.)

The body has been opened on the ventral side by a longitudinal incision extending the whole length of the animal in the right ventral interradius. The body walls and viscera are spread symmetrically on both sides. In the crown of tentacles (*t*) are seen the two smaller ventral tentacles (t^1). From the calcareous ring beneath the tentacles, the slender esophagus (*e*) leads into the saclike stomach (*st*), and thence into the intestine. The intestinal convolutions are not represented in exactly their natural positions, but it is possible to distinguish the three main sections (*i. 1, i. 2, and i. 3*) of this canal. The intestine is attached to the body walls by strong mesenteries, of which only one (*mes*), connected with the posterior section of the intestine, is represented. The cloaca (*cl*) is connected with the body walls by strong muscles (*cl. m*); its connection with the exterior through the cloacal aperture (*cl. ap*) is shown. From the anterior end of the cloaca the pair of profusely branched respiratory trees (*r. t*) extend nearly to the anterior end of the body cavity. The slender filaments of the sexual gland (*g*) lead to a narrow genital duct (*g. d*), which opens on the dorsal side of the body between the bases of the dorsal tentacles. The five pairs of powerful longitudinal muscles (*l. m.*), which mark the five radii of the body, connect anteriorly with the equally strong retractor muscles (*r. m*) of the peristome. At the base of the calcareous ring (*c. r*), beneath the tentacles, the ring canal (*r. c*) of the water-vascular system is seen. This canal is connected with the large saclike Polian vesicles (*P. v*), and the slender madreporic canal (*s. c*), the latter ending in a small madreporic plate.

open into the anterior end of the broad, muscular cloaca. The intestine is supported by three mesenteries, attached to the corresponding interradii. Thus the proximal section is attached along the greater part of the length of the mid-dorsal interradius, the second section to the left dorsal interradius, and the last section to the right ventral interradius, as shown in Fig. 20. In its total length the intestine describes a partial spiral turn around the main axis of the body, as is easily seen by the position of the attachment of the mesenteries.

The mesentery attaching the third section to the right ventral interradius is shown on Plate XXX.*

Calcareous Ring.— Surrounding the anterior end of the esophagus, and forming a support for the oral tentacles, is a broad calcareous ring, consisting of five radial and five interradial plates. The radial plates are notched anteriorly for the passage of the radial nerve and have two long posterior projections, while the interradial plates are much simpler.

Cloaca.— The cloaca serves not only for the discharge of waste substances from the alimentary canal, but also has a most important respiratory function as described below. It is in sharp contrast with the extremely thin walled intestine in that it is provided with very powerful circular and radiating muscles. The former surround the cloaca closely, while the latter branch out radially and pass directly through the cœlom in all directions to connect the cloaca with the adjacent body walls (Plate XXX). The contraction of these radiating muscles causes the cloaca to open widely, while the circular muscles serve to close it. When the cloaca is opened, water rushes in from the outside through the cloacal aperture, and is thus forced up into the respiratory trees.

The cloacal opening is supported by a narrow ring of calcareous plates, with five posterior projections, termed "anal

* Since the animal is represented as opened along the ventral side, the organs on the left side of the body are seen on the right-hand side of the figure. The fact that the mesentery attaching the posterior portion of the intestine to the right ventral interradius appears on the right-hand side of the figure shows that the cut has been made along the animal's *right ventral* interradius, between the mesentery and the right ventral radius. The radius which extends along the median axis of the figure is therefore the left dorsal, while the right dorsal and right ventral are seen on the left-hand side, and the left ventral and ventral on the right-hand side of the figure (compare Fig. 20).

teeth", situated in the five radii. There are also five groups of slender, sensory papillæ surrounding the opening.

Respiratory Trees.*— The anterior end of the cloaca leads forward into a broad sac, which divides into a pair of very thin-walled tubes, the so-called respiratory trees. These are profusely branched (Plate XXX), and extend forward nearly to the anterior end of the body. Their very numerous broad-lobed ramifications (which are made too slender in the figure) expose a vast amount of surface to the cœlomic fluids.

The musculature described above enables the cloaca to be widely opened and filled with the external sea water. By closing the cloacal aperture this water is forced up into the respiratory trees under considerable pressure. In this way pure water is supplied to the branches of the respiratory trees, and because of the pressure a portion of the water is forced through the walls of the branches and enters the cœlom. This cavity is thus constantly filled with water, and the body walls kept tense and firm.

Such a continuous supply is necessary to replace the loss which constantly occurs by diffusion through the tentacles and tube-feet during the active movements of the animal.

By constantly pumping the water in and out of the cloaca the body receives a constant supply of pure water, bringing in the oxygen necessary for life, and carrying out the waste formed in the metabolic processes.

When the animal is disturbed and contracts its body violently, at the same time withdrawing its tentacles, the water in the respiratory trees is driven out of the cloacal aperture with such force that it forms a jet which may be ejected for a considerable distance.

In *Synapta* the respiratory trees are wanting. Instead, there are numerous ciliated funnels attached to the left dorsal mesentery which probably have an excretory function.

Water-vascular System.†— The ring canal at the base of the tentacles (Plate XXX) connects with radial canals which extend to the posterior end of the body between the two bands of

---

* By injecting colored starch paste through the cloaca, the respiratory trees may be made to show all their branches with great distinctness.

† It is possible to inject the water-vascular system through one of the Polian vesicles with a colored fluid, such as Prussian blue solution. By this method the radial canals with their branches may be rendered quite conspicuous.

longitudinal muscles in each of the five radii. Connected with the ring canal is the madreporic canal, which, strangely enough, ends in a madreporic plate situated in the midst of the cœlomic cavity, as shown in Plate XXX. The water of this cavity is thus used for the supply of the water-vascular system. Connected with the ring canal are the large Polian vesicles, which vary in number in different individuals from one to two, or even three. When three are present, one of them is usually quite rudimentary. They contain a reserve supply of fluid.

Scattered all over the internal surface of the body are the slender ampullæ connected with the tube-feet, or pedicels.

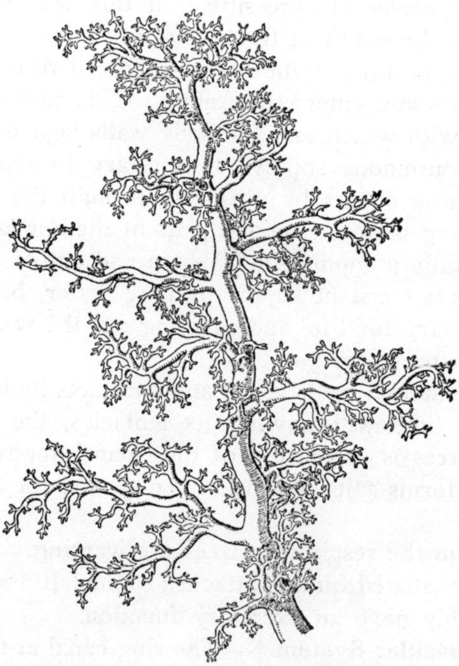

FIG. 21. *Thyone briareus.* One of the oral tentacles, showing the profusion of branches.

Both the very numerous pedicels, which occur in the inter-radial areas as well as in the radii, and the large, branched tentacles are connected with the radial canals of the water-vascular system. Both oral tentacles (Fig. 21) and pedicels are provided with muscular ampullæ, which receive the fluid when the pedicel or tentacle is contracted, as in the starfish.

In *Synapta,* in which pedicels are wanting, the water-vascular system is very rudimentary, and no radial canals occur, the tentacles being connected with the ring canal.

Musculature.— Paired bands of powerful longitudinal muscles extend the length of the body in each of the five radii, while transverse, or circular, muscles extend between the adjacent radii. In all except *Synapta* the circular muscles are interrupted at the radii.

Attached to the calcareous ring are powerful retractor muscles by means of which the crown of tentacles, and in fact the whole anterior part of the body, can be completely withdrawn. They are connected with the five pairs of longitudinal muscles well back toward the middle of the body (Plate XXX). In some specimens these retractors are split for the greater portion of their length into pairs, corresponding with those of the longitudinal muscles, but in other individuals the two parts of each muscle are firmly united.

The musculature of the cloaca is described above.

Body Walls.— In this species the walls of the body are so thick, tough and leathery that calcareous plates are wanting except at the anterior and posterior ends of the body. In all our other holothurians, however, characteristic plates are found in all parts of the walls. These plates, as well as those that are found in the tentacles and pedicels, are described in detail in the descriptions of the different species, for they are in each species of such characteristic form that they furnish the most reliable means of recognizing specific differences.

Nervous System.— The quick and vigorous response which the animal makes to various stimuli indicates a highly sensitive nervous system.

In its morphological structure, however, the nervous system, with its nerve ring and radial nerves, is closely similar to that described for the starfish.

Reproductive Organs.— The sexes are separate. The sexual gland is attached to the dorsal mesentery, and consists of a brush-like structure, composed of numerous slender filaments all converging to a central tube, the genital duct. In reality these filaments are in two clusters, one on each side of the mesentery, but are commonly so closely interwoven as to appear as a single cluster, as represented in Plate XXX. The genital duct passes forward as a slender canal in the dorsal interradius. It opens to the exterior in the mid-dorsal line between the bases of the dorsal tentacles. A small papilla marks the opening of this duct in the males.

HABITS AND FOOD

The holothurians are usually very sluggish in their movements, although they respond suddenly and vigorously to various stimuli. Our native species live largely in the mud or sand between tides or on the sea bottom, moving slowly from place to place as conditions of food or other circumstance may require.

In the worm-like *Synapta,* locomotion is accomplished by the contractions of tentacles and body walls, while the other species are provided with very numerous pedicels or tube-feet by means of which they can creep about.

The species of *Synapta* live with the posterior end buried deeply in the sand or mud. In our other holothurians, however, the cloacal opening is connected with the respiratory apparatus. Hence in these species the body is usually bent into a U-shape with both ends near the surface of the sand or mud in which the animals live.

Thyone.—The habits of the common *Thyone* and its response to stimuli have recently been studied and described by Pearse.* Very little is known regarding the natural history of *Thyone unisemita* or *Thyone scabra,* for of these species very few specimens have ever been seen alive.

Living, active individuals of the common *Thyone* kept in an aquarium are most interesting creatures to watch. The delicately branched, purple oral tentacles (Plate XXXI) are constantly being alternately stretched out and withdrawn, the finer branches gliding over the materials on which the body is resting. If imbedded in mud or sand the tentacles spread out widely exploring the surface in all directions. By this means they come in

* *Biological Bulletin,* 1908.

contact with any particles of organic substance or small organisms which may be present. Such materials may then be grasped by one of the tentacles, which is then withdrawn into the mouth, and the adhering particles collected are scraped off to serve as food. The tentacle is then thrust out for further exploration. In such a manner quantities of mud containing more or less nutritive substance pass into the alimentary canal.

When living in sand, however, the animal is able to select the food materials and reject most of the particles of sand. Even when sand particles are conveyed into the mouth they are often brought out again when the tentacle returns in search of more food. The tentacles take their turn in conveying the food, and as a rule only a single tentacle enters the mouth at one time. Another tentacle may bring its catch near the mouth and wait for the first tentacle to be withdrawn before entering. An examination of the stomach contents shows the food to consist of protozoans, small worms, crustacea and other animals, together with diatoms, unicellular and filamentous algæ and fragments of other plants and animals.

The respiratory movements involve the whole body, the body walls expanding and contracting as the water is taken in and expelled from the cloacal opening. The discharged water often forms a jet of some force.

In its natural position, the common *Thyone* lies with the cloacal aperture widely opened and slightly elevated above the surface of the sand or mud. Through this opening water is alternately pumped into the cloaca and the respiratory trees and forcibly expelled. This act of respiration often occurs with a fairly definite rhythm, at intervals varying under different conditions from 15 seconds to more than a minute. The anterior end of the animal may also lie at the surface of the mud, with the main portion of the U-shaped body concealed beneath the surface. From time to time the animal buries also the anterior end and remains for hours with only the tip of the cloaca projecting. If disturbed, there is a sudden jet of water forced from the cloaca, and the end of the body is withdrawn beneath the surface. Even a shadow falling upon the sensory papillæ surrounding the cloacal opening is sufficient to cause this reaction.

If placed upon the surface of the sand the animals may either crawl about or bury themselves. They do not bore their way into the sand by means of the tentacles and anterior end of the body, as is described below for *Synapta,* but lie with the ventral side upon the surface of the sand, and, by alternately contracting and extending their bodies, push the sand aside and gradually sink beneath the surface. The pedicels of the ventral side of the body aid in pushing aside the particles of sand and drawing the body down.

Cucumaria.— According to observations of Grave* at Beaufort, North Carolina, *Cucumaria pulcherrima* likewise lives in a U-shaped tube near the surface of the mud near low-water mark. The individuals of this species even when very young show a remarkable sensitiveness to stimuli caused by vibrations, and quickly withdraw the tentacles at the slightest disturbance. Specimens reared from the free-swimming stage, however, and kept in the laboratory for two or more years lose the irritability characteristic of newly caught individuals, so that even when removed from the aquarium with such force as to tear off many of the tube-feet they may fail to draw in the tentacles.

When feeding, they extend the branched tentacles to half the length of the body and gather up the diatoms and other organic matter from the surface of the mud, as described for *Thyone.* In cold weather they cease feeding and remain for long periods of time with the tentacles withdrawn.

Synapta.— The common *Synapta* secures its food in the same way as *Thyone.*

In life the tentacles are almost constantly in action, reaching out in all directions to explore surrounding objects, the digits, or slender branches of the tentacles, being thrust out and curled up incessantly. They secure the food and serve in burrowing. In making the burrow, as described by Clark,† the animal bores vertically down into the sand with the oral end foremost. The tentacles are used to remove the particles of sand in front.

" Synaptas burrow into the sand head first, and almost always go straight downward for some distance, but when once completely buried, they turn in any direction up or down or on the

* Tentacle Reflex of a Holothurian, *Cucumaria pulcherrima.* *Johns Hopkins Univ. Circulars,* No. 5, pp. 24-27, 1905.

† *Bull. U. S. Fish Commission,* page 25, 1899.

horizontal plane. They can and do turn in their burrows, but as a rule they make new tubes when they come to the surface. They are seldom still, and the old idea that they remained in the tube they have formed with their tentacles just above the surface is scarcely true. Sometimes they assume that position, but seldom remain so very long. They rarely leave their burrows and come out on the surface of the sand, and I doubt if they ever do so under normal conditions.

"Passage through the sand is chiefly accomplished by means of contractions and extensions of the body, but is materially assisted by the tentacles. With the latter, which are almost continually in motion, the sand is loosened and the grains more or less separated. By the contraction of the longitudinal muscles the rear of the body is brought up nearer to the head, and then the circular muscles contract and extend the body again. It is prevented from slipping back by the anchors, which are elevated by the contraction of the circular muscles and hold against the sand. Since the contraction begins next to the rear end and moves forward, the head end is pushed onward, the anchors there lying flat in the skin.

"This process of alternate contractions of the two sets of muscles is very obvious to an observer, and takes place very continuously, though not rapidly. In this way a *Synapta* can move through the sand from 2 to 3 centimeters a minute, and an *inhærens* of average size can get entirely out of sight in 5 or 6 minutes. One of the most remarkable provisions for the use of the anchors in locomotion is *their much greater abundance and their considerably greater length in the posterior part of the body.* The use of this is clear when one realizes how the rear of the body acts as the resisting base against which the muscles work in pushing the anterior end forward."

AUTOTOMY

Both *Thyone* and *Synapta* often exhibit the habit of self-mutilation or autotomy when subjected to very unfavorable conditions or strong stimuli. Such reactions, however, occur in very different ways in the two genera. Merely placing a *Synapta* in a jar of clean sea water without sufficient sand to enable it to bury itself will lead in a short time to violent contractions of the posterior end of the body, resulting in the spasmodic contraction of the circular muscles at certain points with such violence as to divide the body into two or more parts. The anterior part may continue to cut off successive pieces until only a small fragment of the anterior end of the body remains. This consists of the tentacles, calcareous ring, nerve ring, and circular canal of the water-vascular system, yet this fragment can under favorable conditions restore the normal proportions of the body.

The posterior fragments, severed from the nerve ring, may become much constricted but do not suffer further fragmentation. Such parts are unable to regenerate the whole body.

In *Thyone,* the mutilation consists in such powerful contractions of the body muscles as to cause the rupture of the body walls and longitudinal muscles immediately behind the calcareous ring. This results in the ejection of the nerve ring, circular canal of the water-vascular system, and a portion of the alimentary canal, together with the whole crown of tentacles. Pearse* describes a series of experiments to determine the conditions under which such autotomy occurs. Various chemical stimuli and foul water are the most effective agents in producing it. Such mutilation as occurs in *Thyone* is doubtless entirely pathological and probably causes the death of the animal in nearly all cases. Pearse makes the statement, however, that regeneration of the lost parts may ultimately occur.

REGENERATION

In their regenerative processes some of the holothurians exhibit powers far beyond those of any other animals of such a high degree of specialization. When disturbed, certain species contract their bodies with such violence as to force out the greater portion of their viscera, including not only the cloaca and respiratory apparatus, but also nearly the whole of the alimentary canal. Under favorable conditions the normal body is restored. While none of our own species have either this habit or its compensating regenerative faculty, they are able to restore their normal proportions after very considerable mutilation.

---

* *Biol. Bulletin,* vol. xviii, pp. 42-49, 1909.

The following interesting observations by Clark,* show the ability of the common *Synapta* to regenerate missing parts.

"A number of experiments were made to test the tenacity of life and the possibility of regeration in synaptas, and the results show that *inhærens* is not a very sensitive animal. If an individual is cut in two the anterior end will live and grow as well apparently as any normal specimen, but the posterior end will only live for a few hours, or perhaps a day. So far as I could see, the only reason for its death was its inability to take in food; and I am inclined to think that if food could be provided the posterior half would live as well as the anterior. It seemed to make no difference whether the bisection occurred near to the head or far from it; the head always lived, and in the course of two weeks would show perceptible signs of growth. All that seemed to be necessary was the mouth and a small part of the digestive tract. That it was not the tentacles which were essential was shown by the fact that synaptas lived all right without them. Two or more tentacles were cut from a number of synaptas — in two cases every one being removed — yet they all lived and burrowed in the sand with more or less ease; and not only did they live, but regeneration began at once, so that in two weeks the new tentacles were large enough to bear a digit on each side. At first I thought the nerve ring was the essential part, but that seems to be doubtful, for the nerve ring was carefully and completely severed in a synapta without apparently causing any inconvenience. In other specimens it was cut in two or even three places, but with the same result. In no case were any serious effects shown, and the animals lived and burrowed in the sand with apparently as much ease as ever. I made no microscopical examination to show whether the nerve ends reunited or not, but the cuts apparently healed in a short time. The most striking fact was that with the cutting of the nerve ring there was not the least evidence of any lack of coördination in the movements of the tentacles, nor of the muscles of the body."

---

* *Bull. U. S. Fish Commission*, 1899.

REPRODUCTION AND DEVELOPMENT

In the holothurians, as in the other groups of echinoderms, the sexes are usually separate. Most species of *Synapta,* however, are hermaphroditic, the sexual glands forming both eggs and spermatozoa, which mature at different periods.

The eggs of our common species are beautifully transparent, and have a very symmetrical type of cleavage, but are not so easy to fertilize artificially as are those of the starfish or sea-urchin.

Although three of the species inhabiting Long Island Sound are found in great abundance in certain localities, very little is known regarding their development. The absence of such knowledge seems the more remarkable when we consider the fact that the eggs of all three species are matured in summer, when the opportunities for investigating problems in marine biology are most favorable.

Grave* has been successful in rearing the free-swimming larvæ of *Cucumaria pulcherrima* from the fertilized eggs. He has also observed the transformations to the adult form and has even been able to keep the young holothurians alive in the laboratory for two or three years. This has been accomplished by furnishing the animals with an abundant supply of diatoms and other minute organisms growing at the surface of a layer of sand. The method consists in introducing into the aquarium a small quantity of the surface sand which has been dredged from the ocean bottom and allowed to stand in a jar of sea water for several days. The surface will then be found to be covered with a film of diatoms. The aquarium with the larvæ to be reared, after receiving a small quantity of this diatom culture, is covered and placed before a window where it will be well lighted but not exposed to the direct rays of the sun. The diatoms grow, and furnish an abundant supply of the natural food of the animals, and at the same time keep up a constant supply of oxygen. The water may be changed every few weeks to replace the salts used by the diatoms and echinoderms.†

As in other groups, the eggs of the holothurians in which the development is known, give rise to free-swimming embryos, which

---

* *Johns Hopkins Univ. Circulars,* No. 5, pp. 24-27, 1905.

† C. Grave. A Method of Rearing Marine Larvæ. *Science,* vol. xv, pp. 579-580, 1902.

are at first somewhat similar to those of the starfish, and are termed *auriculariæ*. In the later stages, the embryos of many species become barrel-shaped.

After swimming and feeding at the surface for several weeks the embryo settles to the bottom and transforms itself into a young holothurian.

In a few species from other parts of the world, the young develop within the body cavity of the parent.

KEY TO SPECIES

The class Holothuroidea is conveniently divided into two orders, based on the characters of the water-vascular system and the distribution of pedicels:— (1) the *Actinopoda*, in which the radial canals of the water-vascular system are well developed and supply the oral tentacles, and (2) the *Paractinopoda*, in which the radial canals and pedicels are absent, the oral tentacles being supplied from the circumoral canal. The first order includes our species of *Cucumaria* and *Thyone*, and the second order includes the two species of *Synapta*.

The following key will serve to distinguish our native species:

1. Pedicels present; body not very slender.................2
   Pedicels wanting; body slender and worm-like............5
2. Pedicels scattered over the whole body..................3
   Pedicels mainly limited to the five ambulacral regions; body whitish ....**Cucumaria pulcherrima**
3. Ventral ambulacral area conspicuously demarcated by a distinct border free from pedicels; color white, yellowish white, or flesh-color................**Thyone unisemita**
   Ventral ambulacral area not conspicuously demarcated; pedicels very numerous all over body; color brown or black 4
4. Body tapering posteriorly, but not attenuated; pedicels delicate; few calcareous plates in body wall. **Thyone briareus**
   Body attenuated posteriorly; pedicels and body wall rough from closely crowded calcareous plates...**Thyone scabra**
5. Color white or yellowish; the five radial plates of calcareous ring perforated for passage of nerve..**Synapta inhærens**
   Color reddish or pink; radial plates of calcareous ring notched in anterior border............**Synapta roseola**

Cucumaria pulcherrima (Ayres)

Plate XXXII, figs. 2 to 6.

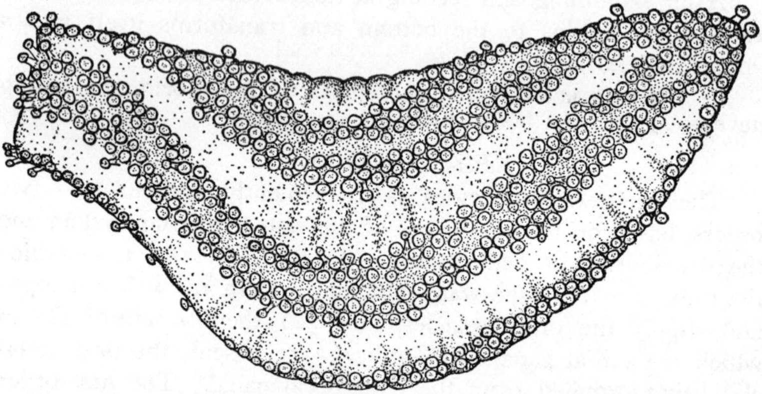

FIG. 22. *Cucumaria pulcherrima.* Specimen preserved in alcohol, showing ambulacral areas and arrangement of tube-feet (pedicels). Five times natural size.

This pretty little holothurian, formerly known as *Pentamera pulcherrima,* has been collected at many localities from Vineyard Sound to South Carolina, although very few specimens have been taken from their natural habitat. Living animals are rarely found even in systematic dredging. In certain localities, however, considerable numbers are washed up on the beach after hard storms. Most of the specimens in collections have been secured in this way. The species has been dredged at least once in Vineyard Sound, although in very thorough dredgings in the same region during several subsequent years no specimens were secured. In the very locality where large numbers of individuals have been washed up on the beach during the winter, systematic search by means of dredgings on the bottoms off shore the following summer were without success. It has been found on the beach after severe storms at several localities on the Connecticut shore. Among the seaweed thrown up by the waves on the rocky points both east and west of New Haven, the species has been found repeatedly during the colder months of the year.

In the harbor of Beaufort, North Carolina, the species occurs near low-water mark in muddy localities where there is a growth of eelgrass. Grave has reared the larvæ to the adult form, and has succeeded in keeping the young holothurians to the age of two or three years by the methods described on page 128.

Color white or pale yellowish. Size usually about 1 inch long and a half-inch in diameter; rarely reaching a length of 2 inches.

In such specimens as are washed up by the waves the body is oval, sometimes tapering towards one end; with a distinct curve, the two ends being bent dorsally, with the result that the length of the body measured along the ventral surface is much greater than if measured along the dorsal surface.

There are ten much branched tentacles, of which the two ventral are much smaller than the others, as in *Thyone*. In the specimens found on the beach they are usually fully retracted into the body, as shown in Plate XXXII, figs. 2 to 6.

The tube-feet, or pedicels, are rather stout, and are strictly limited to the five ambulacral regions, or radii (Fig. 22).

The cloacal opening is surrounded by five small groups of pedicels which are stiffened with thickly placed calcareous deposits (Fig. 23). Imbedded in the body walls in all parts of

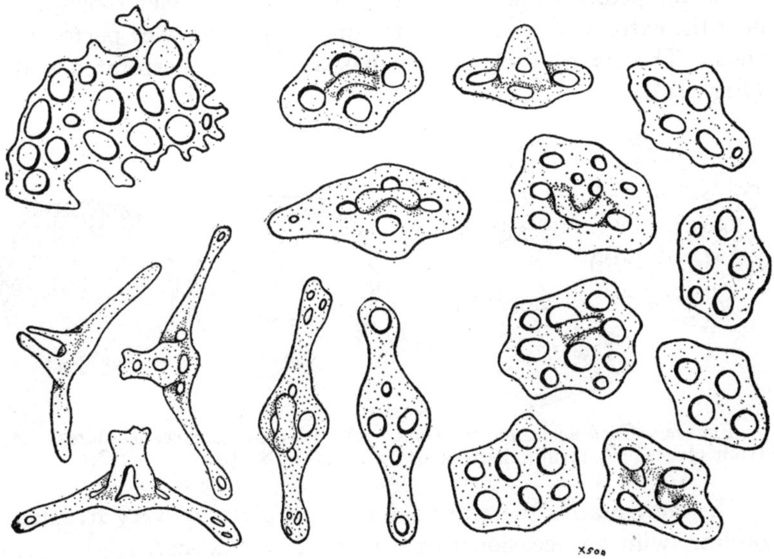

FIG. 23. *Cucumaria pulcherrima.* Outlines of calcareous plates. The upper left-hand plate is from a tentacle; the three lower left-hand plates are from the pedicels; all the others are from the body wall. (× 300.)

the body are closely crowded calcareous deposits of characteristic form (Fig. 23). They greatly increase the firmness of the integument. Some of these deposits are quite symmetrical, while others are very irregular. The most typical shape is that of a table in form of a parallelogram, perforated by four oval holes, and surmounted by a spire consisting of two rods with a few sharp points at the apex (Fig. 23). Others are rectangular, with six holes, and a small spire. Still others have eight, ten, or occasionally more holes. Some are considerably elongated, and, when unusually slender, have one or two small holes in each extremity. The vast majority have graceful, curved or wavy outlines, and are remarkably symmetrical. At the posterior end of the body the tables become densely packed together, and are of much larger size than those found elsewhere. Near the cloaca the spire of the tables becomes much reduced, the deposits assuming the form of large perforated plates.

The cloacal opening is supported by a ring of crowded calcareous plates of large size, but quite irregular in shape (Fig. 24).

In the pedicels the calcareous deposits are slender tables, and near the extremity are reduced to supporting rods with perforated ends. The tip of each pedicel has a terminal perforated plate (Fig. 24).

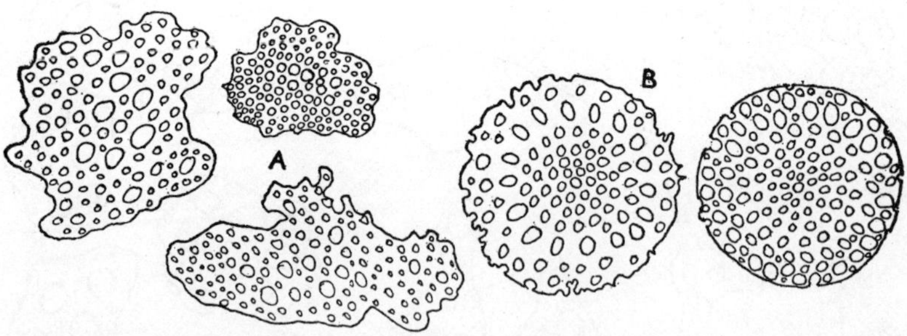

FIG. 24. *Cucumaria pulcherrima.* Outlines of calcareous plates. *A*, from cloaca; *B*, terminal disks of pedicels. (× 150.)

In the tentacles there are a few small plates of very irregular outline, with an occasional one of much larger size.

PLATE XXXI. Common Thyone, *Thyone briareus.*

Thyone briareus (Leseur)

Common Thyone

Plate XXIX, fig. 5; Plates XXX and XXXI.

This species is distributed from Vineyard Sound to the Gulf of Mexico, occurring from near low-water mark to a depth of ten fathoms on both sandy and muddy bottoms. In certain localities it is very abundant, while in seemingly equally favorable situations it does not occur. It is commonly found buried in soft mud in shallow water, often with its body bent into a U-shape, with both extremities at the surface of the mud. In this position its much branched, purplish tentacles stretch out on the surface of the mud in search of the fine particles of organic matter which constitute its food. The posterior end, with the cloacal aperture usually widely opened, likewise reaches the surface of the mud and alternately sucks in and discharges the water used in respiration. The species is often found at the bottom of streams and creeks where the water is decidedly brackish.

The color of the body is purplish, brownish, or blackish, the pedicels lighter, often with a reddish cast, and the terminal disks yellowish.

The body when mature may become six or even nine inches in length when fully extended, most individuals, however, being about three or four inches long and an inch or more in diameter. In shape, the animal is largest in its middle portions, being somewhat narrower anteriorly, and tapering decidedly toward the posterior end (Plate XXXI); but the whole body may become regularly oval or much rounded in certain states of contraction.

The mouth, situated in the center of the anterior end, is surrounded by a crown of ten tentacles, of which the two ventral are much smaller than the others. All are profusely branched, as shown in Fig. 21. The bases of these tentacles and their larger branches are stiffened by the development of numerous

---

Explanation of Plate XXXI. *Thyone briareus.*

Fig. 1. Portion of one of the oral tentacles, showing the terminal finger-like processes. Much enlarged.

Fig. 2. Large specimen with oral tentacles expanded, and pedicels fully extended. Natural size.

Fig. 3. Preserved specimen with oral tentacles partially contracted. Natural size.

perforated calcareous plates, and somewhat similar plates occur more sparingly in the smaller branches. They are either rod-like or broad, and are commonly perforated by six or more holes (Fig. 25).

The body walls, especially in mature specimens, are without calcareous deposits in the greater part of the body. At both extremities, however, there are numerous peculiar calcareous deposits known as "tables." These tables are more or less nearly square or rectangular, and are perforated by about eight rounded holes. At right angles to this plate is the so-called spire, made up of about four short rods ending in teeth and connected by one or two cross processes (Fig. 25).

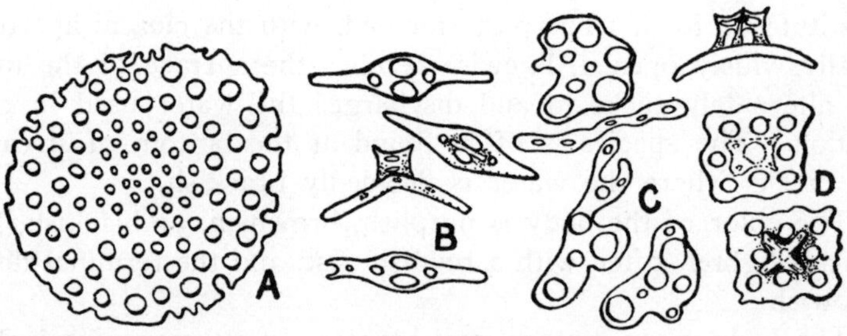

FIG. 25. Outlines of calcareous plates of *Thyone briareus*. *A*, terminal plate of pedicel; *B*, supporting plates of pedicels; *C*, plates from tentacles; *D*, plates (tables) from body-walls.

The slender pedicels, or tube-feet, are very numerous, and are scattered irregularly all over the body. Occasionally they appear to be arranged in five very broad longitudinal bands, although such bands are not distinctly separated from each other. The pedicels of the ventral surface are broader than those of the dorsal side of the body, the latter commonly tapering to a point. In the extremity of the pedicel is a terminal circular plate (Fig. 25), perforated by many minute holes. In the walls of the pedicels are other perforated plates, some of which are modified to form supporting plates, while others are provided with a short spire (Fig. 25).

The cloacal opening has five groups of small papillæ, and a calcareous ring with the five tooth-like processes known as the anal teeth.

The calcareous ring at the base of the tentacles is very broad, and the radial plates have rather slender prolongations.

The internal anatomy of this species is described in detail on pages 116 to 122, and is illustrated on Plate XXX. The habits and reactions to stimuli have recently been studied by Pearse, and are described in an interesting article in the *Biological Bulletin,* Vol. XV, pp. 259-288, 1908. The sexual products mature in June.

Thyone unisemita (Stimpson)

Single-striped Thyone

Plate XXXII, fig. 1.

In the earlier writings of Ayres and Verrill this species was placed in the genus *Stereoderma,* and as such appears in Verrill and Smith's Invertebrate Animals of Vineyard Sound. There are few available records of the collection of this species in Long Island Sound. The United States Fish Commission collected twelve specimens off Watch Hill, Rhode Island, at a depth of eleven fathoms. Six of these specimens are preserved in the Yale University Museum. The Fish Commission likewise records the collection of two other specimens in Long Island Sound, but without giving the number of station or precise locality. The species has also been dredged on sandy bottoms at various localities from southern New England northward to the Banks of Newfoundland, but has nowhere been found in abundance.

The living animals are white, flesh-colored, or yellowish white, with orange-yellow tentacles. Most specimens are less than two inches in length, although one individual measuring three inches in length has been recorded.

The body is somewhat U-shaped, and narrowed at both ends. The pedicels cover the dorsal and lateral surfaces, but on the ventral side there is only one double row situated along the mid-ventral line. This row is rendered the more conspicuous because it is bordered on each side by a wide area of smooth skin, giving the characteristic effect of a single ventral stripe.

The calcareous deposits in the body wall (Fig. 26) consist of flattened plates, many of which are symmetrical and perforated by four holes. Others are larger, irregular in outline, and with

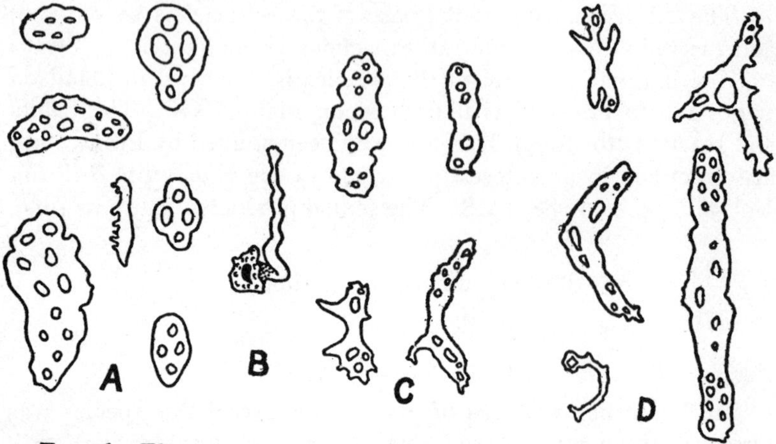

FIG. 26. *Thyone unisemita.* Outlines of calcareous plates. *A,* from body-wall; *B,* madreporic plate; *C,* from pedicels; *D,* from tentacles. (After Clark.)

numerous perforations. The madreporic plate and the characteristic plates of the pedicels and tentacles are shown in Fig. 26.

Further details of the anatomy may be found in Clark's Report on the Echinoderms of the Woods Hole Region.

Thyone scabra Verrill

Rough-skinned Thyone

Plate XXXII, fig. 7.

This species has not as yet been recorded from Connecticut waters. It is, however, included in this report because it has been collected near the eastern entrance to Long Island Sound, and is reported from Narragansett Bay. Hence its occurrence

---

EXPLANATION OF PLATE XXXII.

Fig. 1. *Thyone unisemita.* Three specimens preserved in alcohol, with completely retracted peristome and tentacles. Slightly enlarged.

Figs. 2-6. *Cucumaria pulcherrima.* Five specimens after preservation in formalin, showing variations in shape of body and ambulacral areas. Slightly enlarged.

Fig. 7. *Thyone scabra.* Preserved specimen with extended peristome, and partially extended tentacles. Slightly enlarged.

Fig. 8. *Synapta inhærens.* Anterior portions of body of two individuals with partially contracted tentacles. Natural size.

Fig. 9. *Synapta inhærens.* Posterior portion of body, showing two of the five longitudinal muscular bands. Twice natural size.

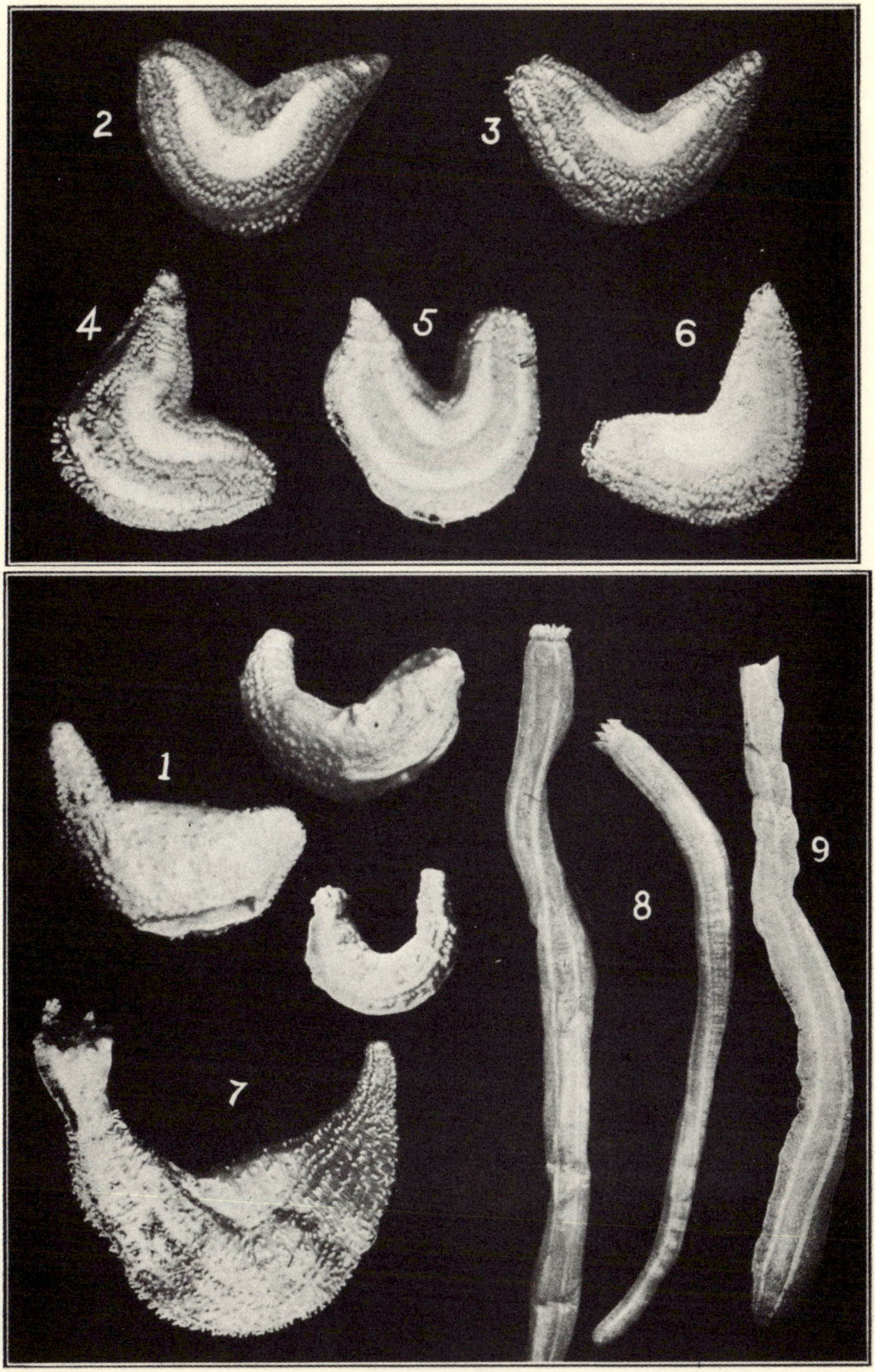

PLATE XXXII. *Thyone, Cucumaria, and Synapta.*

is probable in the colder and deeper waters at the eastern end of the Sound.

The species ranges northward to the Bay of Fundy, and has been reported as far south as Delaware, although the record for the last-named region may not be without question. It extends from shallow water to a depth of more than 600 fathoms.

There are no records as to the color or general appearance of the living animal, the species having been studied and described from preserved specimens. When preserved in alcohol, the color is brownish, with lighter pedicels. Nothing is known regarding its habits.

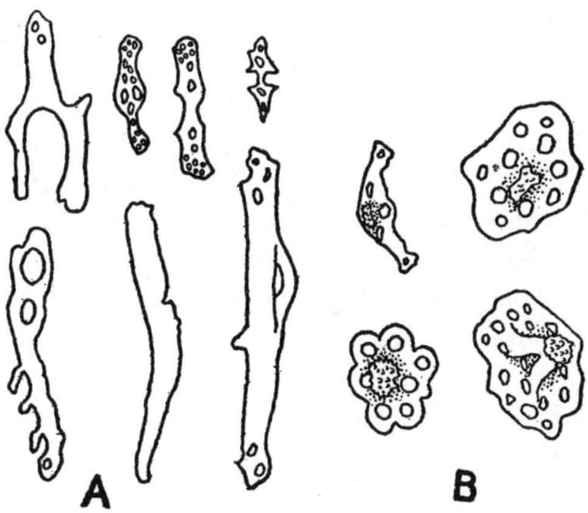

FIG. 27. *Thyone scabra.* Outlines of calcareous plates. *A,* from tentacles; *B.* from body-wall. (After Clark.)

The individuals of this species are very much smaller than those of the common *Thyone,* rarely exceeding two inches in length. The body is slender posteriorly and pointed at the extremity.

The most striking characteristic of the species is the abundance of the calcareous deposits in the skin. These are placed so closely together as to cause a decided roughness over the whole surface of the body. The deposits consist mainly of irregular plates (Fig. 27), usually pierced by eight to twelve or more holes, and provided with a rather small handle, or spire. The spire is made up of two or three stout columns, united at the top and crowned with small, irregular teeth.

The plates in the pedicels are narrow and curved, with four central perforations besides one or more holes at each end; while those of the tentacles are rod-like, often with irregular outline, and perforated (Fig. 27).

For other anatomical details see Clark's Report in the Bulletin of the U. S. Fish Commission for 1902, or Verrill's original description.*

Synapta inhærens (O. F. Müller)

Common Synapta

Plate XXIX, figs. 3, 4; Plate XXXII, figs. 8, 9.

It was formerly thought that the common New England *Synapta* represented a distinct species, to which the name *Leptosynapta girardii* was applied, and this name is used in Verrill and Smith's Report on the Invertebrate Animals of Vineyard Sound. It has been shown, however, that our form is specifically identical with the well-known European *Synapta inhærens.* A very complete description of the anatomy and habits of this species has been published by Clark, in the Bulletin of the U. S. Fish Commission for 1899.

This is a very widely distributed species, being found not only on the Atlantic coast of the United States and along the European shore from the Arctic Ocean to the Mediterranean Sea, but also on the Pacific coast of North America from Alaska to Monterey Bay and Point Loma, California. It also occurs at Bermuda.

Along the Connecticut shore it is very commonly distributed, and occurs in a great variety of situations.

It is most frequently seen on sandy flats between tides, where its burrows can be recognized, when the water is low, by the small mound of sand commonly of different color from that at the surface. In the center of this mound can be seen the small hole in which the creature is living. This hole descends perpendicularly, with a diameter of from one-fourth to three-eighths of an inch.

Although most commonly found in clean sand, it also occurs in muddy situations, and can even live in soft, black mud.

******Am. Journ. Science,* **ser. 3, vol. v, 1873.**

Furthermore it is often found in coarse gravel, and sometimes beneath large stones. And finally, it occurs not only between tides, where the mouths of its burrows may be exposed to the air for an hour or more, but also below low-water mark, extending to a depth of over 100 fathoms. When found in sand between tides its body remains deep in the wet sand beneath the surface, where there is an abundance of water for respiration.

Its ability of adapting itself to such a wide range in the character of its environment, probably accounts for its very wide geographical distribution.

The body is long, slender, cylindrical and worm-like, and capable of great extension. It is so delicate and fragile that complete specimens are difficult to preserve intact. Its length is often 4 to 6 or more inches, with a diameter of ¼ to ⅜ of an inch.

The color of the body is white, often with a distinct tinge of yellow and occasionally of very pale red. The rosy hue of some specimens is due to minute particles of red pigment in the integument, but the reddish color is never so distinct as in the succeeding species. The integument is thin and delicate, and often so translucent that the five bands of longitudinal muscles, which are opaque white in color, and certain of the internal organs, can be distinguished in the living animal. This is particularly true of the intestine, which is filled with sand, and of the reproductive organs when the sexual products are mature.

There are twelve tentacles (Plate XXIX, fig. 3), with characteristic branches. Each tentacle commonly consists of a main stalk and five or seven pairs of branches (Plate XXIX, fig. 4), though there are occasionally only three pairs. The branches are tubular or finger-like projections, and are termed digits. Each tentacle has a number of peculiar sense organs on the inner surface near the base. Calcareous deposits, consisting of oval or knobbed plates or rods, occur in the tentacles.

The most characteristic features of the species of this genus, however, are the very remarkable calcareous deposits found in the body walls. These consist of structures resembling miniature anchors, each of which is connected with a perforated plate of characteristic design (Fig. 28). These occur in all parts of the body,* but are smaller anteriorly than elsewhere, increasing gradually in size toward the posterior end of the body.

The anchor has a narrow shank, with rather short arms; on the outer side of the flukes are several minute spines. The shank is narrowed near the point where it joins the short, slightly curved stock.

The plate, or table, on which the anchor rests has seven symmetrically placed perforations, with serrated margins, followed by a transverse row of three oval perforations, with smooth margins, and several small, smooth holes toward the posterior border. Forming an arch above the three oval perforations is a short

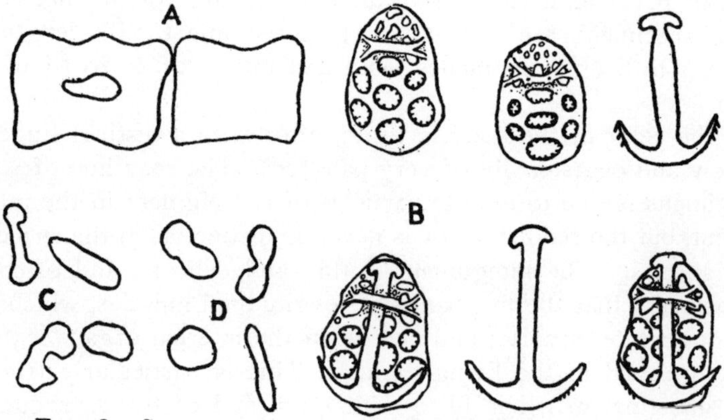

FIG. 28. *Synapta inhærens.* Calcareous plates. *A*, portion of calcareous ring; *B*, anchors and plates from body-wall; *C*, plates from tentacles; *D*, plates from longitudinal muscles. (*A*, *C*, and *D* after Clark.)

transverse process, raised above the level of the plate by a pair of supports at each end (Fig. 28).

The calcareous ring is broad, the plates being without projections on either side. The radial plates are perforated near the center for the passage of the radial nerves. This character alone will serve to distinguish this species from the following. Of the internal anatomy, it is necessary to mention only the large, spiral intestine, which extends the whole length of the body, the single

* These anchors and plates make most beautiful microscopic objects. They can be prepared by soaking a piece of the body wall in a solution of caustic potash until it is thoroughly softened, when it can be washed in water, and the deposits mounted as desired.

Polian vesicle, the terminal cloaca, and the very slender filaments of the gonads. Respiratory trees are absent, and, as the body is without pedicels, the radial vessels of the water-vascular system are wanting in the adult.

In Long Island Sound the animals are sexually mature in the late spring and early summer. The eggs appear fully ripe during the last week in May on the sand flats of Savin Rock. Attempts at artificial fertilization have not been successful.

When removed from its burrow, the animal is very restless, and, if it is roughly handled, or if the water becomes stale, the posterior portion of the body is likely to break into numerous pieces.

This species secures its food by extending its tentacles upon the surface of the sand at the mouth of its burrow, and grasping with the digits such particles of organic substances and sand as may be selected. In this way vast quantities of sand are devoured, the digestible substances contained among the particles or adhering to them being absorbed. The intestine is always filled with sand or mud, which may be of such a color as to affect the coloration of the posterior part of the body.

The presence of the anchors in the integument enables the animals to cling closely to objects with which they may come in contact. It thus often happens that, when specimens of *Synapta* are collected with specimens of other animals, and placed in a jar of water, the whole collection may become entangled into an inextricable mass.

Specimens may be kept alive in an aquarium with a deep layer of clean sand in the bottom, for some weeks, if there is a constant supply of pure sea water.

Synapta roseola (Verrill)

Pink Synapta

Plate XXIX, figs. 1, 2.

This beautiful little holothurian resembles the preceding species closely, but is easily distinguished by the granules of red pigment scattered so thickly over the body as to give it a delicate rose-pink color. It has, moreover, certain anatomical peculiarities by which it can be recognized, even after the color has disappeared from preserved specimens. Individuals of this species are usually smaller than those of *S. inhærens,* fully mature specimens rarely measuring much over 4 inches in length and ¼ inch in diameter.

The body is slender and cylindrical, with a crown of delicate tentacles at the anterior end. It is, as in the preceding species, very fragile. The animal is prone to break up the posterior portion of its body into many pieces when removed from its burrows. The color is pink, pale red, or sometimes fairly deep red, occasionally with a tinge of yellow. The color is mainly due to deep red granules of pigment scattered over the integument, particularly in the anterior portion of the body. When contracted, the color naturally becomes more intense than when extended. When sexually mature the filamentous gonads are conspicuous through the delicate body walls.

This species ranges from Massachusetts Bay to Long Island Sound, and is very abundant at Bermuda. Although it is locally distributed, it is very common in certain places.

The sexual products are ripe in summer, the genital glands at that time filling the anterior portions of the body, and plainly visible through the translucent body walls. Occasionally the eggs may be artificially fertilized and reared to the stage of free-swimming larvæ, but for some unknown reason the majority of attempts at artificial fertilization have been unsuccessful.

Its habits are somewhat different from those of *Synapta inhærens,* for this species is more commonly found in coarse gravel and beneath and between stones. It sometimes occurs in sand. In certain localities specimens may be most easily obtained by turning over the stones beneath which the animals live. It occurs both between tides and below low-water mark.

The anatomical peculiarities which distinguish this species from *Synapta inhærens* are fully discussed by Clark.*

Each of the twelve tentacles has usually but two or three pairs of digits (Plate XXIX, figs. 1 and 2). The calcareous deposits in the tentacles are very irregular, much branched, or perforated, while those in the longitudinal muscles are irregular or incomplete rings (Fig. 29).

---

* The Synaptas of N. E. Coast. *Bull. U. S. Fish Commission,* 1899.

The anchors and plates which are found in all parts of the body wall are very similar to those in *S. inhærens,* but both

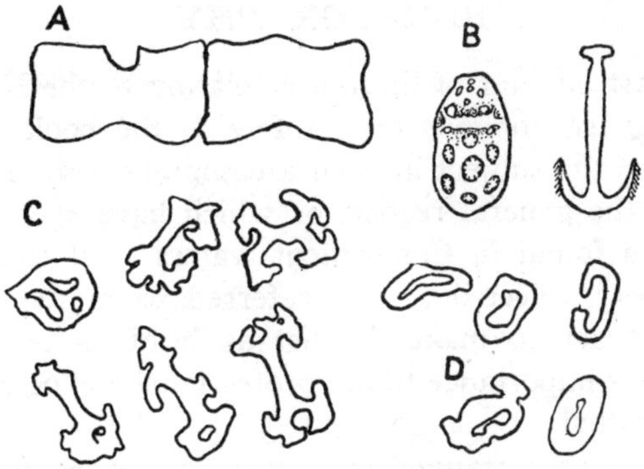

FIG. 29. *Synapta roseola.* Calcareous plates. *A,* portion of calcareous ring; *B,* anchor and plate from body-wall; *C,* plates from tentacles; *D,* plates from longitudinal muscles. (After Clark.)

anchors and plates have a tendency to be more slender (Fig. 29). It is the calcareous ring, however, which furnishes the most precise criterion of distinction, for in this species all the plates are rather narrow, and the radial plates have a notch on the anterior border for the passage of the radial nerve (Fig. 29), instead of the perforation found in *S. inhærens* (Fig. 28).

Both species occur on the sand flats at Savin Rock, near New Haven, in great abundance.

Clark, in his extensive Monograph on the Apodous Holothurians (Smithsonian Contributions to Knowledge, vol. XXXV, 1907), has revised the old genus *Synapta* and now places our two native species, *inhærens* and *roseola,* in Verrill's genus *Leptosynapta.* Although his reasons for doing this appear to be perfectly valid, it has been thought best to retain the long familiar generic name in the present work.

BIBLIOGRAPHY

Of the vast amount of literature relating to the Echinoderms, the following list contains only a few of the books and papers which treat of the subject in such a comprehensive manner as to be useful to the general reader, or which have special reference to the species found in Connecticut waters. Of the latter only such have been selected as are referred to in the text of the article. In order to make the list as brief as possible, many papers of great importance to the professional zoölogist have been purposely omitted.

The articles are arranged in alphabetical order, with the date immediately following the name of the author for convenience of reference.

Agassiz, A. 1872-1874.

Revision of the Echini. *Ill. Cat. Mus. Comp. Zoöl.*, No. 7.

A most important work on the structure and relationships of the echinoids.

Agassiz, A. 1877.

North American Starfishes. *Mem. Mus. Comp. Zoöl.*, Vol. v, No. 1.

A beautifully illustrated monograph on the external anatomy of our native starfishes.

Agassiz, Elizabeth C. and **Alexander.** 1865.

Seaside Studies in Natural History. Marine Animals of Massachusetts Bay. Radiates. Boston. Houghton, Mifflin and Company.

This little book contains a popular account of some of our most interesting marine invertebrates, including many of the echinoderms.

Arnold, Augusta Foote. 1901.

The Sea Beach at Ebb Tide. New York City. The Century Company.

A well illustrated and interestingly written guide to the study of the marine invertebrates and plants.

Ayres, W. O. 1851-1854.

Notices of Holothuriæ and other Echinoderms. *Proc. Bost. Soc. Nat. Hist.*, Vol iv.

Although written more than a half-century ago, the descriptions of the natural history of many of our native species are as interesting as any that have been published since.

Bather, F. A., assisted by **Gregory, J. W.**, and **Goodrich, E. C.** 1900.

The Echinoderma. Lankester's Treatise on Zoölogy, Pt. iii.

One of the most valuable works on the general anatomy, relationships, and natural history of the group.

Brooks, W. K.

Handbook of Invertebrate Zoölogy.

A well-known guide to the laboratory study of invertebrates.

Clark, H. L. 1899.

The Synaptas of the New England Coast. *Bull. U. S. Fish Commission.*

A clearly written and well illustrated report on our two native species of *Synapta.*

Clark, H. L. 1901.

Synopsis of North American Invertebrates. XV. The Holothuroidea. *American Naturalist,* Vol. xxxv, 1901.

A brief account of the anatomical peculiarities of the holothurians, followed by analytical keys to the genera and species found on both the Atlantic and Pacific coasts of North America.

Clark, H. L. 1904.

The Echinoderms of the Woods Hole Region. *Bull. U. S. Fish Commission* for 1902.

Concise systematic descriptions of nearly all the species of echinoderms found in the waters of southern New England. Each species is illustrated by excellent photographs, most of which were made from living animals.

Fewkes, J. W. 1887.

On the Development of the Calcareous Plates of *Amphiura. Bull. Mus. Comp. Zoöl.*, Vol. xiii, No. 4.

An important contribution to the embryology of the ophiurans.

Fisher, W. K. 1911.

Asteroidea of the North Pacific and adjacent waters. Part 1, Phanerozonia and Spinulosa. Bull. 76, *U. S. National Museum.*

An extensive monograph, containing detailed anatomical descriptions and a critical discussion of the relationships of the starfishes of the North Pacific. Illustrated by 122 quarto plates.

Ludwig, H. 1892.

Die Seewalzen. Bronn's Tierreich, Band ii.

A classic monograph on the holothurians.

Lyman, T. 1865.

Ophiuridæ and Astrophytidæ. *Ill. Cat. Mus. Comp. Zööl.,* No. 1.

Contains a systematic account of the American ophiurans.

Lyman, T. 1882.

Report on the Ophiuroidea. Challenger Reports, Vol. v, Pt. 14.

A systematic treatise on the ophiurans of the world.

Mac Bride, E. W. 1906.

Echinodermata. Cambridge Natural History.

One of the most important general accounts of the echinoderms ever written. Contains many excellent illustrations of the anatomical structures and developmental stages of the various groups.

Mead, A. D. 1900.

The Natural History of the Starfish. *Bull. U. S. Fish Commission* for 1899, Vol. xix.

The most interesting account of the natural history of the common starfish.

Preyer, W. 1887.

Ueber die Bewegungen der Seesterne. *Mitt. aus der Zoöl. Station zu Neapel,* Band 7.

A classical paper on the movements of starfish.

Romanes, G. J. 1885.

Jellyfish, Starfish, and Sea-urchins. London. International Scientific Series.

A most entertaining popular account of the natural history of these groups of animals.

Sladen, W. P. 1889.
Report on the Asteroidea. Challenger Reports, Vols. xxx and xxxi.
A monograph of the starfishes of the world.

Theel, H. 1886.
Report on the Holothurioidea. Challenger Reports, Vol. xiv.
A monograph of the holothurians of the world.

von Uexküll, J. 1909.
Umwelt und Innenwelt der Tiere. Berlin.
A series of most interesting stories about the habits of representatives of various groups of invertebrates, including the classes of echinoderms, with an account of their behavior under a great variety of natural and artificial conditions.

Verrill, A. E., and **Smith, S. I.** 1874.
Report on the Invertebrate Animals of Vineyard Sound and Adjacent Waters. *Report of Commissioner of Fish and Fisheries,* Vol. i, for 1871-2. Washington.
The most comprehensive report on the marine invertebrates of southern New England that has yet been published. Although some new species have been discovered since the work appeared, and the scientific names of some of the species have been revised, it is still the standard work on the subject. The accounts of the natural history of the very numerous species included in the report are most interestingly written.

Verrill, A. E. 1895.
Distribution of the Echinoderms of Northeastern America. *Am. Journ. Sci.,* Vol. xlix.
Discusses the geographical and bathymetrical distribution of the various species.

Verrill, A. E. 1899.
North American Ophiuroidea. *Trans. Conn. Acad.,* Vol. x.
Revision of the synonymy and distribution of many of the ophiurans of the east coast of North America and the West Indies.

INDEX

A CATALOGUE OF SELECTED DOVER BOOKS IN ALL FIELDS OF INTEREST

A CATALOGUE OF SELECTED DOVER BOOKS IN ALL FIELDS OF INTEREST

AMERICA'S OLD MASTERS, James T. Flexner. Four men emerged unexpectedly from provincial 18th century America to leadership in European art: Benjamin West, J. S. Copley, C. R. Peale, Gilbert Stuart. Brilliant coverage of lives and contributions. Revised, 1967 edition. 69 plates. 365pp. of text.
21806-6 Paperbound $3.00

FIRST FLOWERS OF OUR WILDERNESS: AMERICAN PAINTING, THE COLONIAL PERIOD, James T. Flexner. Painters, and regional painting traditions from earliest Colonial times up to the emergence of Copley, West and Peale Sr., Foster, Gustavus Hesselius, Feke, John Smibert and many anonymous painters in the primitive manner. Engaging presentation, with 162 illustrations. xxii + 368pp.
22180-6 Paperbound $3.50

THE LIGHT OF DISTANT SKIES: AMERICAN PAINTING, 1760-1835, James T. Flexner. The great generation of early American painters goes to Europe to learn and to teach: West, Copley, Gilbert Stuart and others. Allston, Trumbull, Morse; also contemporary American painters—primitives, derivatives, academics—who remained in America. 102 illustrations. xiii + 306pp. 22179-2 Paperbound $3.00

A HISTORY OF THE RISE AND PROGRESS OF THE ARTS OF DESIGN IN THE UNITED STATES, William Dunlap. Much the richest mine of information on early American painters, sculptors, architects, engravers, miniaturists, etc. The only source of information for scores of artists, the major primary source for many others. Unabridged reprint of rare original 1834 edition, with new introduction by James T. Flexner, and 394 new illustrations. Edited by Rita Weiss. 6⅝ x 9⅝.
21695-0, 21696-9, 21697-7 Three volumes, Paperbound $13.50

EPOCHS OF CHINESE AND JAPANESE ART, Ernest F. Fenollosa. From primitive Chinese art to the 20th century, thorough history, explanation of every important art period and form, including Japanese woodcuts; main stress on China and Japan, but Tibet, Korea also included. Still unexcelled for its detailed, rich coverage of cultural background, aesthetic elements, diffusion studies, particularly of the historical period. 2nd, 1913 edition. 242 illustrations. lii + 439pp. of text.
20364-6, 20365-4 Two volumes, Paperbound $6.00

THE GENTLE ART OF MAKING ENEMIES, James A. M. Whistler. Greatest wit of his day deflates Oscar Wilde, Ruskin, Swinburne; strikes back at inane critics, exhibitions, art journalism; aesthetics of impressionist revolution in most striking form. Highly readable classic by great painter. Reproduction of edition designed by Whistler. Introduction by Alfred Werner. xxxvi + 334pp.
21875-9 Paperbound $2.50

VISUAL ILLUSIONS: THEIR CAUSES, CHARACTERISTICS, AND APPLICATIONS, Matthew Luckiesh. Thorough description and discussion of optical illusion, geometric and perspective, particularly; size and shape distortions, illusions of color, of motion; natural illusions; use of illusion in art and magic, industry, etc. Most useful today with op art, also for classical art. Scores of effects illustrated. Introduction by William H. Ittleson. 100 illustrations. xxi + 252pp.
21530-X Paperbound **$2.00**

A HANDBOOK OF ANATOMY FOR ART STUDENTS, Arthur Thomson. Thorough, virtually exhaustive coverage of skeletal structure, musculature, etc. Full text, supplemented by anatomical diagrams and drawings and by photographs of undraped figures. Unique in its comparison of male and female forms, pointing out differences of contour, texture, form. 211 figures, 40 drawings, 86 photographs. xx + 459pp. 5⅜ x 8⅜.
21163-0 Paperbound $3.50

150 MASTERPIECES OF DRAWING, Selected by Anthony Toney. Full page reproductions of drawings from the early 16th to the end of the 18th century, all beautifully reproduced: Rembrandt, Michelangelo, Dürer, Fragonard, Urs, Graf, Wouwerman, many others. First-rate browsing book, model book for artists. xviii + 150pp. 8⅜ x 11¼.
21032-4 Paperbound **$2.50**

THE LATER WORK OF AUBREY BEARDSLEY, Aubrey Beardsley. Exotic, erotic, ironic masterpieces in full maturity: Comedy Ballet, Venus and Tannhauser, Pierrot, Lysistrata, Rape of the Lock, Savoy material, Ali Baba, Volpone, etc. This material revolutionized the art world, and is still powerful, fresh, brilliant. With *The Early Work,* all Beardsley's finest work. 174 plates, 2 in color. xiv + 176pp. 8⅛ x 11.
21817-1 Paperbound **$3.00**

DRAWINGS OF REMBRANDT, Rembrandt van Rijn. Complete reproduction of fabulously rare edition by Lippmann and Hofstede de Groot, completely reedited, updated, improved by Prof. Seymour Slive, Fogg Museum. Portraits, Biblical sketches, landscapes, Oriental types, nudes, episodes from classical mythology—All Rembrandt's fertile genius. Also selection of drawings by his pupils and followers. "Stunning volumes," *Saturday Review*. 550 illustrations. lxxviii + 552pp. 9⅛ x 12¼.
21485-0, 21486-9 Two volumes, Paperbound $7.00

THE DISASTERS OF WAR, Francisco Goya. One of the masterpieces of Western civilization—83 etchings that record Goya's shattering, bitter reaction to the Napoleonic war that swept through Spain after the insurrection of 1808 and to war in general. Reprint of the first edition, with three additional plates from Boston's Museum of Fine Arts. All plates facsimile size. Introduction by Philip Hofer, Fogg Museum. v + 97pp. 9⅜ x 8¼.
21872-4 Paperbound $2.00

GRAPHIC WORKS OF ODILON REDON. Largest collection of Redon's graphic works ever assembled: 172 lithographs, 28 etchings and engravings, 9 drawings. These include some of his most famous works. All the plates from *Odilon Redon: oeuvre graphique complet,* plus additional plates. New introduction and caption translations by Alfred Werner. 209 illustrations. xxvii + 209pp. 9⅛ x 12¼.
21966-8 Paperbound $4.00

A History of Costume, Carl Köhler. Definitive history, based on surviving pieces of clothing primarily, and paintings, statues, etc. secondarily. Highly readable text, supplemented by 594 illustrations of costumes of the ancient Mediterranean peoples, Greece and Rome, the Teutonic prehistoric period; costumes of the Middle Ages, Renaissance, Baroque, 18th and 19th centuries. Clear, measured patterns are provided for many clothing articles. Approach is practical throughout. Enlarged by Emma von Sichart. 464pp. 21030-8 Paperbound $3.50

Oriental Rugs, Antique and Modern, Walter A. Hawley. A complete and authoritative treatise on the Oriental rug—where they are made, by whom and how, designs and symbols, characteristics in detail of the six major groups, how to distinguish them and how to buy them. Detailed technical data is provided on periods, weaves, warps, wefts, textures, sides, ends and knots, although no technical background is required for an understanding. 11 color plates, 80 halftones, 4 maps. vi + 320pp. 6⅛ x 9⅛. 22366-3 Paperbound $5.00

Ten Books on Architecture, Vitruvius. By any standards the most important book on architecture ever written. Early Roman discussion of aesthetics of building, construction methods, orders, sites, and every other aspect of architecture has inspired, instructed architecture for about 2,000 years. Stands behind Palladio, Michelangelo, Bramante, Wren, countless others. Definitive Morris H. Morgan translation. 68 illustrations. xii + 331pp. 20645-9 Paperbound $2.50

The Four Books of Architecture, Andrea Palladio. Translated into every major Western European language in the two centuries following its publication in 1570, this has been one of the most influential books in the history of architecture. Complete reprint of the 1738 Isaac Ware edition. New introduction by Adolf Placzek, Columbia Univ. 216 plates. xxii + 110pp. of text. 9½ x 12¾. 21308-0 Clothbound $10.00

Sticks and Stones: A Study of American Architecture and Civilization, Lewis Mumford.One of the great classics of American cultural history. American architecture from the medieval-inspired earliest forms to the early 20th century; evolution of structure and style, and reciprocal influences on environment. 21 photographic illustrations. 238pp. 20202-X Paperbound $2.00

The American Builder's Companion, Asher Benjamin. The most widely used early 19th century architectural style and source book, for colonial up into Greek Revival periods. Extensive development of geometry of carpentering, construction of sashes, frames, doors, stairs; plans and elevations of domestic and other buildings. Hundreds of thousands of houses were built according to this book, now invaluable to historians, architects, restorers, etc. 1827 edition. 59 plates. 114pp. 7⅞ x 10¾. 22236-5 Paperbound $3.00

Dutch Houses in the Hudson Valley Before 1776, Helen Wilkinson Reynolds. The standard survey of the Dutch colonial house and outbuildings, with constructional features, decoration, and local history associated with individual homesteads. Introduction by Franklin D. Roosevelt. Map. 150 illustrations. 469pp. 6⅝ x 9¼. 21469-9 Paperbound $4.00

JOHANN SEBASTIAN BACH, Philipp Spitta. One of the great classics of musicology, this definitive analysis of Bach's music (and life) has never been surpassed. Lucid, nontechnical analyses of hundreds of pieces (30 pages devoted to St. Matthew Passion, 26 to B Minor Mass). Also includes major analysis of 18th-century music. 450 musical examples. 40-page musical supplement. Total of xx + 1799pp.
(EUK) 22278-0, 22279-9 Two volumes, Clothbound $15.00

MOZART AND HIS PIANO CONCERTOS, Cuthbert Girdlestone. The only full-length study of an important area of Mozart's creativity. Provides detailed analyses of all 23 concertos, traces inspirational sources. 417 musical examples. Second edition. 509pp. (USO) 21271-8 Paperbound $3.50

THE PERFECT WAGNERITE: A COMMENTARY ON THE NIBLUNG'S RING, George Bernard Shaw. Brilliant and still relevant criticism in remarkable essays on Wagner's Ring cycle, Shaw's ideas on political and social ideology behind the plots, role of Leitmotifs, vocal requisites, etc. Prefaces. xxi + 136pp.
21707-8 Paperbound $1.50

DON GIOVANNI, W. A. Mozart. Complete libretto, modern English translation; biographies of composer and librettist; accounts of early performances and critical reaction. Lavishly illustrated. All the material you need to understand and appreciate this great work. Dover Opera Guide and Libretto Series; translated and introduced by Ellen Bleiler. 92 illustrations. 209pp.
21134-7 Paperbound **$1.50**

HIGH FIDELITY SYSTEMS: A LAYMAN'S GUIDE, Roy F. Allison. All the basic information you need for setting up your own audio system: high fidelity and stereo record players, tape records, F.M. Connections, adjusting tone arm, cartridge, checking needle alignment, positioning speakers, phasing speakers, adjusting hums, trouble-shooting, maintenance, and similar topics. Enlarged 1965 edition. More than 50 charts, diagrams, photos. iv + 91pp. 21514-8 Paperbound $1.25

REPRODUCTION OF SOUND, Edgar Villchur. Thorough coverage for laymen of high fidelity systems, reproducing systems in general, needles, amplifiers, preamps, loudspeakers, feedback, explaining physical background. "A rare talent for making technicalities vividly comprehensible," R. Darrell, *High Fidelity.* 69 figures. iv + 92pp. 21515-6 Paperbound **$1.25**

HEAR ME TALKIN' TO YA: THE STORY OF JAZZ AS TOLD BY THE MEN WHO MADE IT, Nat Shapiro and Nat Hentoff. Louis Armstrong, Fats Waller, Jo Jones, Clarence Williams, Billy Holiday, Duke Ellington, Jelly Roll Morton and dozens of other jazz greats tell how it was in Chicago's South Side, New Orleans, depression Harlem and the modern West Coast as jazz was born and grew. xvi + 429pp.
21726-4 Paperbound $2.50

FABLES OF AESOP, translated by Sir Roger L'Estrange. A reproduction of the very rare 1931 Paris edition; a selection of the most interesting fables, together with 50 imaginative drawings by Alexander Calder. v + 128pp. 6½x9¼.
21780-9 Paperbound $1.50

EAST O' THE SUN AND WEST O' THE MOON, George W. Dasent. Considered the best of all translations of these Norwegian folk tales, this collection has been enjoyed by generations of children (and folklorists too). Includes True and Untrue, Why the Sea is Salt, East O' the Sun and West O' the Moon, Why the Bear is Stumpy-Tailed, Boots and the Troll, The Cock and the Hen, Rich Peter the Pedlar, and 52 more. The only edition with all 59 tales. 77 illustrations by Erik Werenskiold and Theodor Kittelsen. xv + 418pp. 22521-6 Paperbound $3.00

GOOPS AND HOW TO BE THEM, Gelett Burgess. Classic of tongue-in-cheek humor, masquerading as etiquette book. 87 verses, twice as many cartoons, show mischievous Goops as they demonstrate to children virtues of table manners, neatness, courtesy, etc. Favorite for generations. viii + 88pp. 6½ x 9¼.
22233-0 Paperbound $1.25

ALICE'S ADVENTURES UNDER GROUND, Lewis Carroll. The first version, quite different from the final *Alice in Wonderland,* printed out by Carroll himself with his own illustrations. Complete facsimile of the "million dollar" manuscript Carroll gave to Alice Liddell in 1864. Introduction by Martin Gardner. viii + 96pp. Title and dedication pages in color. 21482-6 Paperbound $1.25

THE BROWNIES, THEIR BOOK, Palmer Cox. Small as mice, cunning as foxes, exuberant and full of mischief, the Brownies go to the zoo, toy shop, seashore, circus, etc., in 24 verse adventures and 266 illustrations. Long a favorite, since their first appearance in St. Nicholas Magazine. xi + 144pp. 6⅝ x 9¼.
21265-3 Paperbound $1.75

SONGS OF CHILDHOOD, Walter De La Mare. Published (under the pseudonym Walter Ramal) when De La Mare was only 29, this charming collection has long been a favorite children's book. A facsimile of the first edition in paper, the 47 poems capture the simplicity of the nursery rhyme and the ballad, including such lyrics as I Met Eve, Tartary, The Silver Penny. vii + 106pp. 21972-0 Paperbound $1.25

THE COMPLETE NONSENSE OF EDWARD LEAR, Edward Lear. The finest 19th-century humorist-cartoonist in full: all nonsense limericks, zany alphabets, Owl and Pussycat, songs, nonsense botany, and more than 500 illustrations by Lear himself. Edited by Holbrook Jackson. xxix + 287pp. (USO) 20167-8 Paperbound $2.00

BILLY WHISKERS: THE AUTOBIOGRAPHY OF A GOAT, Frances Trego Montgomery. A favorite of children since the early 20th century, here are the escapades of that rambunctious, irresistible and mischievous goat—Billy Whiskers. Much in the spirit of *Peck's Bad Boy,* this is a book that children never tire of reading or hearing. All the original familiar illustrations by W. H. Fry are included: 6 color plates, 18 black and white drawings. 159pp. 22345-0 Paperbound $2.00

MOTHER GOOSE MELODIES. Faithful republication of the fabulously rare Munroe and Francis "copyright 1833" Boston edition—the most important Mother Goose collection, usually referred to as the "original." Familiar rhymes plus many rare ones, with wonderful old woodcut illustrations. Edited by E. F. Bleiler. 128pp. 4½ x 6⅜. 22577-1 Paperbound $1.25

AGAINST THE GRAIN (A REBOURS), Joris K. Huysmans. Filled with weird images, evidences of a bizarre imagination, exotic experiments with hallucinatory drugs, rich tastes and smells and the diversions of its sybarite hero Duc Jean des Esseintes, this classic novel pushed 19th-century literary decadence to its limits. Full unabridged edition. Do not confuse this with abridged editions generally sold. Introduction by Havelock Ellis. xlix + 206pp. 22190-3 Paperbound $2.00

VARIORUM SHAKESPEARE: HAMLET. Edited by Horace H. Furness; a landmark of American scholarship. Exhaustive footnotes and appendices treat all doubtful words and phrases, as well as suggested critical emendations throughout the play's history. First volume contains editor's own text, collated with all Quartos and Folios. Second volume contains full first Quarto, translations of Shakespeare's sources (Belleforest, and Saxo Grammaticus), Der Bestrafte Brudermord, and many essays on critical and historical points of interest by major authorities of past and present. Includes details of staging and costuming over the years. By far the best edition available for serious students of Shakespeare. Total of xx + 905pp.
21004-9, 21005-7, 2 volumes, Paperbound $7.00

A LIFE OF WILLIAM SHAKESPEARE, Sir Sidney Lee. This is the standard life of Shakespeare, summarizing everything known about Shakespeare and his plays. Incredibly rich in material, broad in coverage, clear and judicious, it has served thousands as the best introduction to Shakespeare. 1931 edition. 9 plates. xxix + 792pp. (USO) 21967-4 Paperbound $3.75

MASTERS OF THE DRAMA, John Gassner. Most comprehensive history of the drama in print, covering every tradition from Greeks to modern Europe and America, including India, Far East, etc. Covers more than 800 dramatists, 2000 plays, with biographical material, plot summaries, theatre history, criticism, etc. "Best of its kind in English," *New Republic.* 77 illustrations. xxii + 890pp.
20100-7 Clothbound $8.50

THE EVOLUTION OF THE ENGLISH LANGUAGE, George McKnight. The growth of English, from the 14th century to the present. Unusual, non-technical account presents basic information in very interesting form: sound shifts, change in grammar and syntax, vocabulary growth, similar topics. Abundantly illustrated with quotations. Formerly *Modern English in the Making.* xii + 590pp.
21932-1 Paperbound $3.50

AN ETYMOLOGICAL DICTIONARY OF MODERN ENGLISH, Ernest Weekley. Fullest, richest work of its sort, by foremost British lexicographer. Detailed word histories, including many colloquial and archaic words; extensive quotations. Do not confuse this with the Concise Etymological Dictionary, which is much abridged. Total of xxvii + 830pp. 6½ x 9¼.
21873-2, 21874-0 Two volumes, Paperbound $6.00

FLATLAND: A ROMANCE OF MANY DIMENSIONS, E. A. Abbott. Classic of science-fiction explores ramifications of life in a two-dimensional world, and what happens when a three-dimensional being intrudes. Amusing reading, but also useful as introduction to thought about hyperspace. Introduction by Banesh Hoffmann. 16 illustrations. xx + 103pp. 20001-9 Paperbound $1.00

POEMS OF ANNE BRADSTREET, edited with an introduction by Robert Hutchinson. A new selection of poems by America's first poet and perhaps the first significant woman poet in the English language. 48 poems display her development in works of considerable variety—love poems, domestic poems, religious meditations, formal elegies, "quaternions," etc. Notes, bibliography. viii + 222pp.
22160-1 Paperbound $2.00

THREE GOTHIC NOVELS: THE CASTLE OF OTRANTO BY HORACE WALPOLE; VATHEK BY WILLIAM BECKFORD; THE VAMPYRE BY JOHN POLIDORI, WITH FRAGMENT OF A NOVEL BY LORD BYRON, edited by E. F. Bleiler. The first Gothic novel, by Walpole; the finest Oriental tale in English, by Beckford; powerful Romantic supernatural story in versions by Polidori and Byron. All extremely important in history of literature; all still exciting, packed with supernatural thrills, ghosts, haunted castles, magic, etc. xl + 291pp.
21232-7 Paperbound $2.00

THE BEST TALES OF HOFFMANN, E. T. A. Hoffmann. 10 of Hoffmann's most important stories, in modern re-editings of standard translations: Nutcracker and the King of Mice, Signor Formica, Automata, The Sandman, Rath Krespel, The Golden Flowerpot, Master Martin the Cooper, The Mines of Falun, The King's Betrothed, A New Year's Eve Adventure. 7 illustrations by Hoffmann. Edited by E. F. Bleiler. xxxix + 419pp. 21793-0 Paperbound $2.50

GHOST AND HORROR STORIES OF AMBROSE BIERCE, Ambrose Bierce. 23 strikingly modern stories of the horrors latent in the human mind: The Eyes of the Panther, The Damned Thing, An Occurrence at Owl Creek Bridge, An Inhabitant of Carcosa, etc., plus the dream-essay, Visions of the Night. Edited by E. F. Bleiler. xxii + 199pp. 20767-6 Paperbound $1.50

BEST GHOST STORIES OF J. S. LEFANU, J. Sheridan LeFanu. Finest stories by Victorian master often considered greatest supernatural writer of all. Carmilla, Green Tea, The Haunted Baronet, The Familiar, and 12 others. Most never before available in the U. S. A. Edited by E. F. Bleiler. 8 illustrations from Victorian publications. xvii + 467pp. 20415-4 Paperbound $3.00

THE TIME STREAM, THE GREATEST ADVENTURE, AND THE PURPLE SAPPHIRE—THREE SCIENCE FICTION NOVELS, John Taine (Eric Temple Bell). Great American mathematician was also foremost science fiction novelist of the 1920's. *The Time Stream,* one of all-time classics, uses concepts of circular time; *The Greatest Adventure,* incredibly ancient biological experiments from Antarctica threaten to escape; The *Purple Sapphire,* superscience, lost races in Central Tibet, survivors of the Great Race. 4 illustrations by Frank R. Paul. v + 532pp.
21180-0 Paperbound $3.00

SEVEN SCIENCE FICTION NOVELS, H. G. Wells. The standard collection of the great novels. Complete, unabridged. *First Men in the Moon, Island of Dr. Moreau, War of the Worlds, Food of the Gods, Invisible Man, Time Machine, In the Days of the Comet.* Not only science fiction fans, but every educated person owes it to himself to read these novels. 1015pp. 20264-X Clothbound $5.00

THE RED FAIRY BOOK, Andrew Lang. Lang's color fairy books have long been children's favorites. This volume includes Rapunzel, Jack and the Bean-stalk and 35 other stories, familiar and unfamiliar. 4 plates, 93 illustrations x + 367pp.
21673-X Paperbound $2.50

THE BLUE FAIRY BOOK, Andrew Lang. Lang's tales come from all countries and all times. Here are 37 tales from Grimm, the Arabian Nights, Greek Mythology, and other fascinating sources. 8 plates, 130 illustrations. xi + 390pp.
21437-0 Paperbound $2.50

HOUSEHOLD STORIES BY THE BROTHERS GRIMM. Classic English-language edition of the well-known tales — Rumpelstiltskin, Snow White, Hansel and Gretel, The Twelve Brothers, Faithful John, Rapunzel, Tom Thumb (52 stories in all). Translated into simple, straightforward English by Lucy Crane. Ornamented with headpieces, vignettes, elaborate decorative initials and a dozen full-page illustrations by Walter Crane. x + 269pp. 21080-4 Paperbound $2.50

THE MERRY ADVENTURES OF ROBIN HOOD, Howard Pyle. The finest modern versions of the traditional ballads and tales about the great English outlaw. Howard Pyle's complete prose version, with every word, every illustration of the first edition. Do not confuse this facsimile of the original (1883) with modern editions that change text or illustrations. 23 plates plus many page decorations. xxii + 296pp.
22043-5 Paperbound $2.50

THE STORY OF KING ARTHUR AND HIS KNIGHTS, Howard Pyle. The finest children's version of the life of King Arthur; brilliantly retold by Pyle, with 48 of his most imaginative illustrations. xviii + 313pp. 6⅛ x 9¼.
21445-1 Paperbound $2.50

THE WONDERFUL WIZARD OF OZ, L. Frank Baum. America's finest children's book in facsimile of first edition with all Denslow illustrations in full color. The edition a child should have. Introduction by Martin Gardner. 23 color plates, scores of drawings. iv + 267pp. 20691-2 Paperbound $2.25

THE MARVELOUS LAND OF OZ, L. Frank Baum. The second Oz book, every bit as imaginative as the Wizard. The hero is a boy named Tip, but the Scarecrow and the Tin Woodman are back, as is the Oz magic. 16 color plates, 120 drawings by John R. Neill. 287pp. 20692-0 Paperbound $2.50

THE MAGICAL MONARCH OF MO, L. Frank Baum. Remarkable adventures in a land even stranger than Oz. The best of Baum's books not in the Oz series. 15 color plates and dozens of drawings by Frank Verbeck. xviii + 237pp.
21892-9 Paperbound $2.00

THE BAD CHILD'S BOOK OF BEASTS, MORE BEASTS FOR WORSE CHILDREN, A MORAL ALPHABET, Hilaire Belloc. Three complete humor classics in one volume. Be kind to the frog, and do not call him names . . . and 28 other whimsical animals. Familiar favorites and some not so well known. Illustrated by Basil Blackwell. 156pp. (USO) 20749-8 Paperbound $1.25

AMERICAN FOOD AND GAME FISHES, David S. Jordan and Barton W. Evermann. Definitive source of information, detailed and accurate enough to enable the sportsman and nature lover to identify conclusively some 1,000 species and sub-species of North American fish, sought for food or sport. Coverage of range, physiology, habits, life history, food value. Best methods of capture, interest to the angler, advice on bait, fly-fishing, etc. 338 drawings and photographs. l + 574pp. 6⅝ x 9⅜.
22383-1 Paperbound $4.50

THE FROG BOOK, Mary C. Dickerson. Complete with extensive finding keys, over 300 photographs, and an introduction to the general biology of frogs and toads, this is the classic non-technical study of Northeastern and Central species. 58 species; 290 photographs and 16 color plates. xvii + 253pp.
21973-9 Paperbound $4.00

THE MOTH BOOK: A GUIDE TO THE MOTHS OF NORTH AMERICA, William J. Holland. Classical study, eagerly sought after and used for the past 60 years. Clear identification manual to more than 2,000 different moths, largest manual in existence. General information about moths, capturing, mounting, classifying, etc., followed by species by species descriptions. 263 illustrations plus 48 color plates show almost every species, full size. 1968 edition, preface, nomenclature changes by A. E. Brower. xxiv + 479pp. of text. 6½ x 9¼.
21948-8 Paperbound $5.00

THE SEA-BEACH AT EBB-TIDE, Augusta Foote Arnold. Interested amateur can identify hundreds of marine plants and animals on coasts of North America; marine algae; seaweeds; squids; hermit crabs; horse shoe crabs; shrimps; corals; sea anemones; etc. Species descriptions cover: structure; food; reproductive cycle; size; shape; color; habitat; etc. Over 600 drawings. 85 plates. xii + 490pp.
21949-6 Paperbound $3.50

COMMON BIRD SONGS, Donald J. Borror. 33⅓ 12-inch record presents songs of 60 important birds of the eastern United States. A thorough, serious record which provides several examples for each bird, showing different types of song, individual variations, etc. Inestimable identification aid for birdwatcher. 32-page booklet gives text about birds and songs, with illustration for each bird.
21829-5 Record, book, album. Monaural. $2.75

FADS AND FALLACIES IN THE NAME OF SCIENCE, Martin Gardner. Fair, witty appraisal of cranks and quacks of science: Atlantis, Lemuria, hollow earth, flat earth, Velikovsky, orgone energy, Dianetics, flying saucers, Bridey Murphy, food fads, medical fads, perpetual motion, etc. Formerly "In the Name of Science." x + 363pp.
20394-8 Paperbound $2.00

HOAXES, Curtis D. MacDougall. Exhaustive, unbelievably rich account of great hoaxes: Locke's moon hoax, Shakespearean forgeries, sea serpents, Loch Ness monster, Cardiff giant, John Wilkes Booth's mummy, Disumbrationist school of art, dozens more; also journalism, psychology of hoaxing. 54 illustrations. xi + 338pp.
20465-0 Paperbound $2.75